Data Visualization

S. Margret Anouncia · Hardik A. Gohel ·
Subbiah Vairamuthu
Editors

Data Visualization

Trends and Challenges Toward Multidisciplinary Perception

Springer

Editors
S. Margret Anouncia
Department of Computer Science
and Engineering
Vellore Institute of Technology
Vellore, Tamil Nadu, India

Subbiah Vairamuthu
Department of Computer Science
and Engineering
Vellore Institute of Technology
Vellore, Tamil Nadu, India

Hardik A. Gohel
Department of Computer Science
University of Houston-Victoria
Victoria, TX, USA

ISBN 978-981-15-2284-0 ISBN 978-981-15-2282-6 (eBook)
https://doi.org/10.1007/978-981-15-2282-6

This Springer imprint is published by the registered company Springer Nature Singapore Pte Ltd.
The registered company address is: 152 Beach Road, #21-01/04 Gateway East, Singapore 189721, Singapore

Foreword

Rarely do I see our government client who does not have enough data, especially energy data. It is widely available. Yet, researchers are overwhelmed with it. In my opinion, the most sophisticated data visualization tools today give you the easy, flexible, custom metrics and reports.

I am Director of Research at Florida International University's Applied Research Center (FIU-ARC) and Principal Investigator of a $24M/5-year Cooperative Agreement between the U.S. Department of Energy (DOE) and FIU in support of DOE's environmental restoration program. I am also the Founder and Director of the DOE-FIU Science and Technology Workforce Development Program and Department of Defense Cyber Fellows Program. I know Dr. Hardik A. Gohel, one of the editors of this esteemed creative data visualization book, since 2016. He is a task lead on the Department of Defense (DOD)—Test Resource Management Center (TRMC) funded cybersecurity and data science project.

The title of this book is very much attracted to me, and I would like to highlight some industrial trends. Data visualization is an emerging field in industry, and as yet, it is not well defined as an academic course. We are drowning in data. As per IBM, every day, 2.5 quintillion bytes of data are created. This is the equivalent of 90% of the world's information—created in the last two years alone. Now this is what we call "big data." But where does it come from? Everywhere, from sensors and social media sites to digital images and videos. We have more data than we know what to do with, so it is time now to organize and make sense of it all. This is where data visualization comes into the picture. Data visualization is a big part of a data scientist's jobs. In the early stages of a project, you will often be doing an exploratory data analysis (EDA) to gain some insights into your data. Creating visualizations really helps make things clearer and easier to understand, especially with larger, high-dimensional datasets. Toward the end of your project, it is important to be able to present your final results in a clear, concise, and compelling manner that your audience, who are often non-technical clients, can understand.

I congratulate all authors and editors of this book for their data visualization state-of-the-art research and publications. I look forward to seeing multiple volumes of the book.

Dr. Leonel Lagos
Director of Research
Applied Research Center at Florida
International University
Miami, Florida, USA

Preface

Visualization is one of the most important components of research presentation and communication due to its ability to synthesize large amounts of data into effective graphics. It is easier for the brain to comprehend an image versus words or numbers making effective graphics an especially important part of academic literature. The increasing accessibility and quantity of data require effective ways to analyze and communicate the information that datasets contain in simple, easy-to-understand formats. Data visualization is a collection of two major components. First is data analysis and data presentation.

Many researchers in academia and industry spend their day sifting through data, combining multiple data sources, and finally getting data ready for the moment of truth: seeing it in a data visualization. Data visualizations are the culmination of all data crunching work—they are supposed to take long numeric lists and complicated Key Performance Parameters (KPPs) and present them in intuitive, easy-to-understand way, that is, if you choose the right visualization for your data. The problem is it is often challenging to choose the right visualization for the data people want to show. Do one want to compare values or analyze a trend? What is the best way to visualize one's data so that it is easy to extract insights? Many people stop short there wondering if a chart, graph, or heatmap will best reveal the bottom line at a glance, or worse, default to a simple pie chart because that is what they are most familiar with. But, data visualizations need to clarify the information. Defaulting to the most common visualization can actually lead to a misinterpretation of data.

This book represents an ongoing study, research, investigation, findings, and future directions into the central question: "What is data visualization in multidisciplinary perception?" The motivation is my Cyber Test Automation Monitoring (CTAM) project at Applied Research Center at Florida International University which I worked.

In order to understand this book and its author's origins, it might help readers to understand a little bit about their interest and what their motivations were for doing research in data visualization.

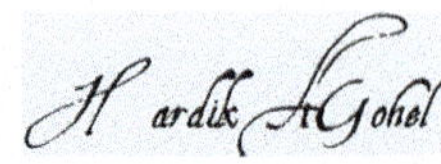

Victoria, USA Hardik A. Gohel, Ph.D.
Vellore, India S. Margret Anouncia
Vellore, India Subbiah Vairamuthu

Contents

About the Editors

Dr. S. Margret Anouncia is a Professor at Vellore Institute of Technology (VIT) University in India. She received her bachelor's degree in Computer Science and Engineering from Bharathidasan University (1993), Tiruchirappalli, India, and master's degree in Software Engineering from Anna University (2000), Chennai, India. She was awarded a doctorate in Computer Science and Engineering at VIT University (2008). Her main areas of interest include digital image processing, software engineering, and knowledge engineering. She is the lead author of more than 60 publications in technical journals and proceedings of national and international conferences.

Dr. Hardik A. Gohel is an assistant professor in computer science and program director of computer information system at University of Houston – Victoria, TX, USA. Postdoctoral Fellow at the Applied Research Center of Florida International University. He worked as a postdoc fellow in cybersecurity and data science at Florida International University – Miami, FL, USA, from 2016–2019. He holds a Ph.D. in Computer Science, awarded in 2015. Dr. Gohel has extensive research experience in artificial intelligence and his research projects have involved cyber test automation and monitoring, smart bandages for wound monitoring, bigdata for security intelligence, trustworthy cyberspace for security and privacy of social media, predictive maintenance for nuclear infrastructure, and database and mobile forensics infrastructure. Dr. Gohel is also working on how to prepare quality diversified workforce with artificial intelligence in science, technology, engineering and mathematics (STEM) education.

Dr. Subbiah Vairamuthu is an Associate Professor at Vellore Institute of Technology (VIT) University in India. He received bachelor's and master's degrees in Computer Science from Ayya Nadar Janaki Ammal College (Autonomous), affiliated to Madurai Kamaraj University. He received his Master of Engineering from Anna University and his Ph.D. from Vellore Institute of Technology. He is the lead author of more than 20 publications in technical journals and proceedings of national and international conferences. His main areas of interest include human–computer interaction, recommendation systems, machine learning, big data and IoT, and computational intelligence.

Narrative and Text Visualization: A Technique to Enhance Teaching Learning Process in Higher Education

S. Anupama Kumar, M. N. Vijayalakshmi, Shaila H. Koppad and Andhe Dharani

Abstract Data visualization is the state-of-the-art technology that enables people from various domains to exhibit their work in a professional way. Visualization helps in understanding things in the past, present and predict the near future. The various tools and techniques of data visualization can be applied to different domains including health care, sales and marketing, forecasting, research, education, etc. This chapter intends the application of two different visualization techniques: narrative and text in the domain of education. Narrative and text visualization are generally considered as two different elements wherein this chapter intends to integrate both the techniques together and make it as a single component to help the education community. Narrative visualization is implemented using a storyboard that converts the text into a narrative format to help the student to understand the course in a simple way. Dashboards are also a narrative visualization technique which helps the student to have a glance of the course material. The content of the text is processed, frequent words are generated and visualized in the form of word cloud and other visualization methods to enable the student to understand and remember the course in a better way as a part of text analytics and visualization. Various techniques and tools including Canva storyboards, R, Tabulae and Power BI are employed in this work to implement narrative and text visualization to enable the tutor to teach better and make the teaching learning process easier.

Keywords Storyboard · Text extraction · Frequent words · Analysis · Word cloud · Stacked bar

1 Introduction

Data visualization is a state-of-the-art technology which enhances people to understand data by providing visual representation. Visualization is the graphical presentation of information, with the goal of providing the viewer with a qualitative understanding of the information contents [1]. It gives the data a completely new

S. Anupama Kumar (✉) · M. N. Vijayalakshmi · S. H. Koppad · A. Dharani
Department of MCA, R V College of Engineering, Bengaluru, India

S. M. Anouncia et al. (eds.), *Data Visualization*,
https://doi.org/10.1007/978-981-15-2282-6_1

meaning and reveals the hidden trends and information present in the data. The various domains ranging from education to research, advertising and marketing, all business set-ups, factories, banking sector and health care make use of data visualization techniques to bring out the new knowledge from it.

Narrative and text visualization are two important visualization techniques which can be applied over various domains to understand data better. Data comes in different formats like text, images and numerals. Narrative visualization techniques convert the data into a storyboard and help us to understand the data better.

Narrative visualization can be implemented by building a storyboard or a dashboard which explains the entire scenario in a better way to the user. Stories are a part of human communication structures, and storytelling enables to pass information and wisdom to the human community in a lively manner. Stories help us make sense of our past and reason about the future. Johnson [2] and Maclntyre [3] argue that story narrative also goes beyond communication; it is also a process of extracting meaning from events that is central to human experience and conduct.

Text visualization techniques are very important in the digital world since the documents are digitized in the current era. Therefore, it becomes essential to adopt a technique which can read the electronic document and help people to easily understand the content of the document without much effort. The visualization technique helps to summarize a document through two main aspects: (1) contents such as words and figures and (2) features such as average sentence length and number of verbs. The contents can further be visualized in the form of word blocks or cloud for explicitly.

The advanced visualization techniques can further create mind maps, visualization trees, blocks and other forms to make the user understand the theme easily. Further topic-based analysis can also be made using specific algorithms, and they can be understood easily.

This chapter intends to integrate narrative and text visualization techniques in the education domain. Narrative visualization will help the tutor to use visualization as a teaching aid to the student and help the student to understand the concepts in a better way. Text visualization will help to create a gist of the course material and help the student to use it as a material and learn the course in an enhanced way.

The following section gives an insight into the coupling of narrative and text visualization techniques.

2 Model for Integrating Narrative and Text Visualization

Narrative visualizations are data visualizations with embedded "stories" presenting particular perspectives using various embedding mechanisms [4]. Kosara and Mackinlay [5] and Lee et al. [6] exhibits storytelling in visualization as a primary way of communication apart from exploratory analysis.

Text visualization can be considered as an information visualization technique of visualizing raw text data or the application of text mining algorithms to visualize textual data [7]. Text visualization can be implemented to visualize the document similarity, revealing the content of the text and depicting the content for easy understanding and usage.

This chapter intends to build a model to couple both narrative and text visualization to bring out various forms of text in a better way and help the tutors to enhance their teaching techniques.

The authors in [8] have coupled the text and narrative visualization manually to interpret the textual data. The authors in [9, 10] analysed text to narrative visualizations using annotations and other elements. The authors in [11] coupled the techniques to implement a technique that can use text documents as well as web links and depict the links in a better way. Different case studies were investigated and proved to ensure the coupling of text and narrative visualization.

In this chapter, the course material (e-book) of a particular course in a higher education system is taken as a data source. Figure 1 shows the coupling of the techniques.

The techniques are implemented as follows:

(i) The course material is available in pdf formats with six chapters. All the chapters are split into separate text documents so that it becomes easy to analyse and visualized in a sophisticated way. The e-book is converted into text files using functions for further process.

(ii) The second step is carried over in two methods.

   a. Storyboards are created using Canva studio. The textual materials in the chapter are converted into visual formats to understand the content in a better manner. Different types of visualizations are used to understand the data in a better way.

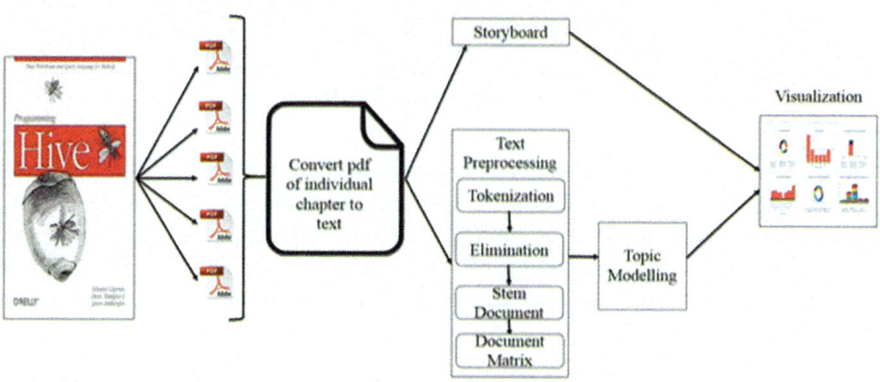

Fig. 1 Coupling of narrative and text visualization

b. The text file is preprocessed to extract relevant information. Then, topic modelling algorithms are implemented to generate frequent topics from each chapter. The above functions are carried out using R.

(iii) The final step is to apply visualization of the applied textual data using appropriate techniques. Narrative visualization is depicted using hierarchical charts, Bubbled charts and dashboards. In case of text visualization, the frequent topics are visualized using different charts/graph using Power BI which is one of the visualization tools.

The following section explains the implementation of narrative and text visualization over the higher educational data.

3 Application of Narrative and Text Visualization

Visualization is one of the most powerful tools that can be implemented over a variety of domains. Education is one of the important domains where visualization helps the tutor to teach better and the learner to understand better. The techniques can be used to enhance the teaching learning process in a better way. The authors in [12] have implemented visualization techniques to teach mathematics for engineering graduates. They have used 3D visualization techniques using JavaScript and WebGL to make the course understandable to learners. The application of visualization in understanding the geo spatial data with multiple dimensions has been well explained in [13]. Joris Klerkx, Katrien Verbert and Erik Duval have discussed the different models, methods and techniques of visualization that can improve and enhance their teaching [14]. They have also discussed various types of learning models and methods over which the visualization techniques can be applied. These methods have given insight to apply narrative and text visualization techniques over the course material.

3.1 Implementation Narrative Visualization

Narrative visualization is one of the effective techniques that can be used for textual representation. But there are few challenges and issues that are to be handled during the process of working. The authors in [15, 16] have listed out various challenges that can arise while implementing narrative visualization over textual data as stories written by authors. The following challenges were met while building a model for narrative visualization

(i) The foremost challenge is inferring the environment in which the events take place. The tutor should be comfortable with the course so that tutor can build the course material in a very sophisticated way.

(ii) Grammatical disfluencies, such as real-word spelling errors, missing punctuation, improper or omitted prepositions, incorrect verb tense and missing auxiliary verbs may cause the syntactic and semantic parsers.
(iii) The narrative text has a very complex theoretical structure which may be complicated for the learner to understand.
(iv) The sequence in which the narration (events) should happen should be depicted properly; otherwise, it may cause disproportionate storyboards to the learner.

Figure 2 shows the raw text material which is taken from the HIVE book of a higher education course.

The objective of this work is to convert this into a meaningful visualization format which will enable the learner to understand the concept easier. The process

CHAPTER 3

Data Types and File Formats

Hive supports many of the *primitive* data types you find in relational databases, as well as three *collection* data types that are rarely found in relational databases, for reasons we'll discuss shortly.

A related concern is how these types are represented in text files, as well as alternatives to text storage that address various performance and other concerns. A unique feature of Hive, compared to most databases, is that it provides great flexibility in how data is encoded in files. Most databases take total control of the data, both how it is persisted to disk and its life cycle. By letting you control all these aspects, Hive makes it easier to manage and process data with a variety of tools.

Primitive Data Types

Hive supports several sizes of integer and floating-point types, a Boolean type, and character strings of arbitrary length. Hive v0.8.0 added types for timestamps and binary fields.

Table 3-1 lists the *primitive* types supported by Hive.

Table 3-1. Primitive data types

Type	Size	Literal syntax examples
TINYINT	1 byte signed integer.	20
SMALLINT	2 byte signed integer.	20
INT	4 byte signed integer.	20
BIGINT	8 byte signed integer.	20
BOOLEAN	Boolean true or false.	TRUE
FLOAT	Single precision floating point.	3.14159
DOUBLE	Double precision floating point.	3.14159

Fig. 2 Raw text material

of creating narrative visualization for course material involves a lot of challenges. The creator of the visualization should be aware of the content of the book so that the visualization can be created accurately. The tutor should consider the following points while creating a storyboard:

(i) The content which the tutor wants to convey should be converted into accurate visualization format which otherwise will convey wrong meaning.
(ii) The important points of the content should be highlighted using appropriate colour, size, boldness and connecting elements like arrows and shaded trails should be properly maintained which otherwise will create trouble to the user.
(iii) The images should be aligned from the largest image to the smallest image with the appropriate caption given to them. The images should also have proper size and should be positioned in appropriate place so that content can be conveyed in proper manner.
(iv) The progress bar and other elements should be properly placed so that it enables the user to understand the importance of the content.
(v) The syntactic and sematic regulations of the content should be properly maintained.
(vi) The space dimensions in the storyboard should be maintained properly to enable the student to visualize the content.

Figure 3 depicts the hierarchical structure created for the given text.

Canva storyboards can be created using various templates provided in the software. The system is user friendly and allows the user to implement the narrative analytics in different ways. The challenge in using these softwares lies about the knowledge of the user who creates the storyboards. The user should also be aware of the scenes that should be created in a sequence. Figure 4 depicts another way of visualizing the same chapter to enable the learner to understand the concept.

The chart depicted in Figs. 3 and 4 will be integrated in the dashboard along with the text visualization to make the learner understand the concept in a better way. The following section explains the implementation of the text processing and topic modelling.

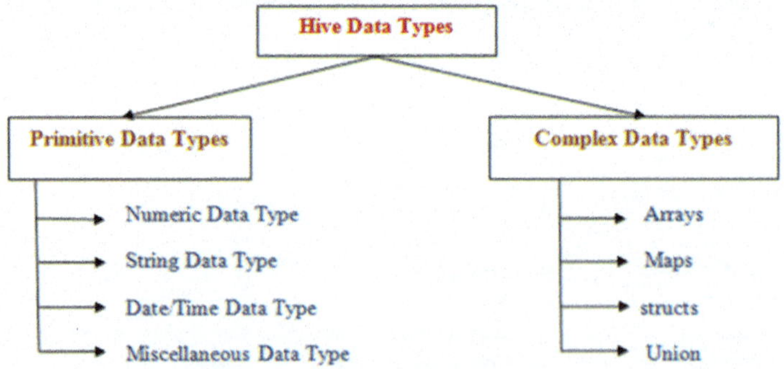

Fig. 3 Hierarchical structure of the raw text

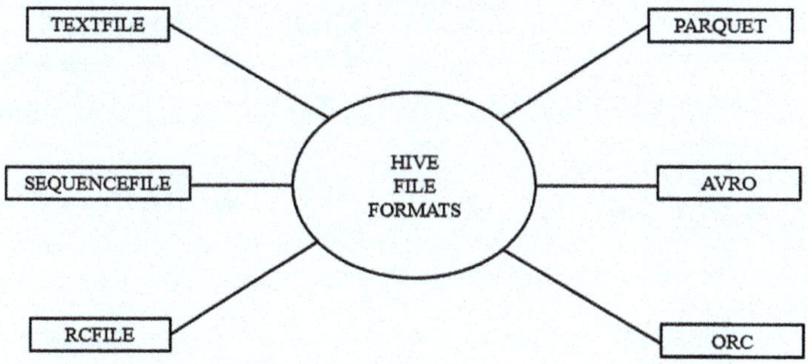

Fig. 4 Narrative text visualization of the chapter

3.2 *Implementation of Text Visualization*

Teaching learning process can be made effective if the learner is able to understand the concept by reading or visualizing the content of the course material in a simple way. Therefore, it becomes the duty of the tutor to make the material interactive and easy for the learner. If the course material is complicated to read and understand, the objective of the course will become diluted. Text processing and topic modelling play a major role in simplifying the course material and provide a gist of the contents to the learner in a better way. In [17], the author has discussed the application of NLP with scientific computer programs to enhance the process of education. Litman [18] has expressed various research areas in education where natural language processing can play a major role and bring in changes in teaching and learning the courses. Natural language processing involves processing the text data (structured or unstructured) and applies text analysis algorithms to get accurate output. The implementation of the text processing and topic modelling is carried using R. The output of topic modelling is depicted in the form of word cloud and other visual aids using R. Figure 5 is the overview of the text visualization process carried over using text processing, topic modelling and visualization.

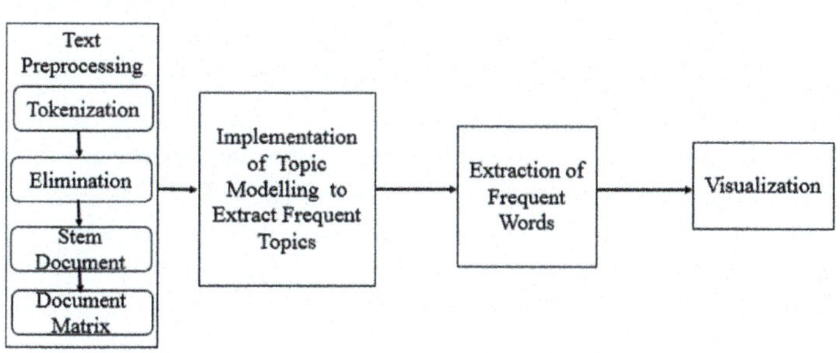

Fig. 5 Overview of text visualization

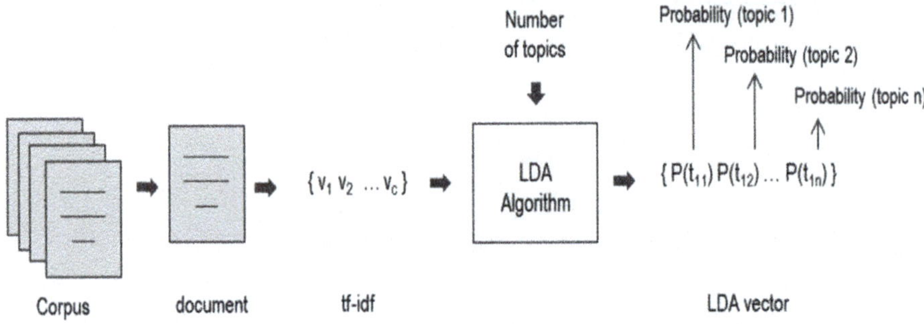

Fig. 6 Implementation of LDA algorithm

Text processing is carried over the course material to extract the raw text from the e-book. This is called as tokenization. Once the words are extracted, it is necessary to remove the stop words, stem the document and filter the document for effective extraction. The numerals and punctuations available in the text are also removed so as to get the plain text. The images and emoticons are not considered in the text processing. A term document matrix is created for each chapter, and the results are recorded.

Topic modelling is one the interesting text mining technique which enables the user to extract topics from a huge corpus. Paul and Girju [19] applied LDA algorithm in interdisciplinary research areas and proved that it can be applied to extract topics from education field also. The working of the LDA algorithm is depicted in Fig. 6.

The course material is converted into the text document and preprocessed in the previous phase. Then, a term document matrix is generated. The term document matrix is fed as an input to the LDA algorithm, and a set of topics are generated depending on the user's demand. The most frequent words from each topic are then extracted from all the chapters. Figure 7 is the resultant sample of frequent words generated using LDA algorithm.

Figure 8 depicts the list of frequency words that are extracted using the algorithm in all the chapters and visualization of common words from the chapter in the form of a cluster. Bar chart is created using Power BI. The most frequent word in the course material is 'hive', the second highest is 'data', and the third is 'Hadoop' and it goes on. The top most words are depicted in green colour in the stacked graph.

The frequent words are imported to Power BI for further visualization. Figure 9 depicts the total number of frequent words in each chapter, and the Bowtie Chart helps to visualize the frequent terms present in the chapter.

The hierarchical structure in Fig. 10 is a sample which shows the visualization of the total frequent words present in the course material and how it is broken down under each chapter.

From Fig. 10, it depicts the occurrence of the words in each chapter. The bubble in the end of first structure states that the process is incomplete, and the words are still to be generated.

hive	69	1	Chapter 1 : Introduction
data	56	1	Chapter 1 : Introduction
hadoop	48	1	Chapter 1 : Introduction
use	41	1	Chapter 1 : Introduction
word	39	1	Chapter 1 : Introduction
mapreduc	33	1	Chapter 1 : Introduction
java	26	1	Chapter 1 : Introduction
count	26	1	Chapter 1 : Introduction
can	25	1	Chapter 1 : Introduction
queri	24	1	Chapter 1 : Introduction
provid	21	1	Chapter 1 : Introduction
languag	21	1	Chapter 1 : Introduction
sql	20	1	Chapter 1 : Introduction
job	18	1	Chapter 1 : Introduction
one	18	1	Chapter 1 : Introduction
key	18	1	Chapter 1 : Introduction

Fig. 7 Sample processed output of LDA algorithm

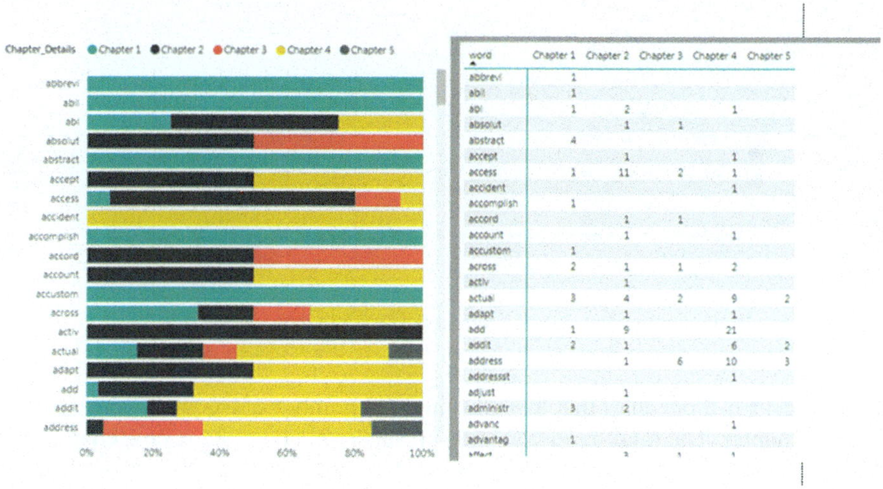

Fig. 8 Frequent words using LDA and its visualization

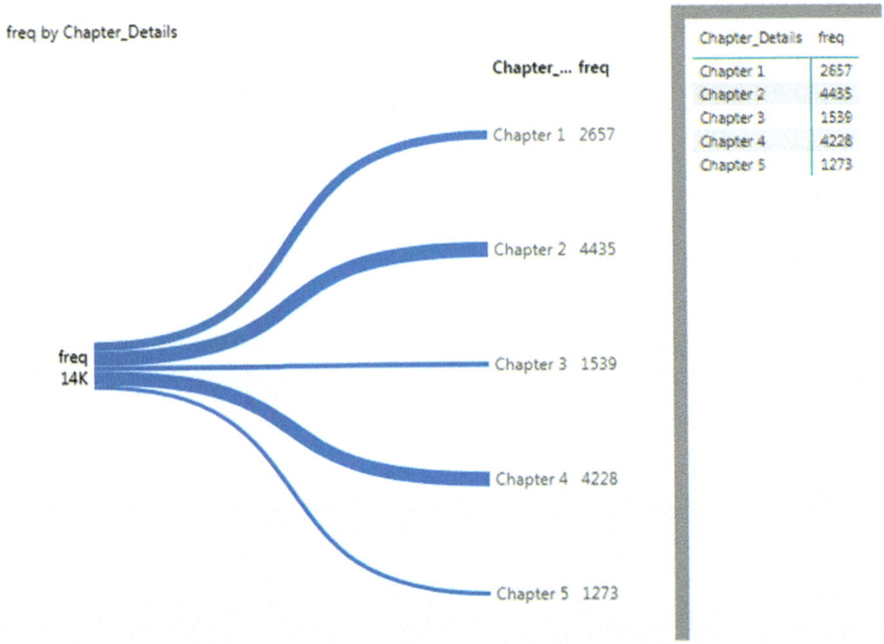

Fig. 9 Bowtie chart representing total frequent words in each chapter

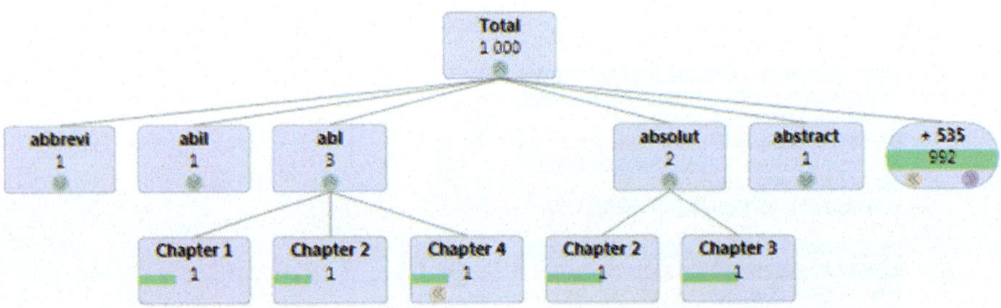

Fig. 10 Visualization of hierarchical structure of frequent words

Figure 11 is a pie chart that explains the total number of frequent words present in each chapter. The total frequent words present in chapter one is 869, chapter two 1003, chapter three 495, chapter four 881 and chapter five 402.

3.3 *Integration of Narrative and Text Visualization*

The final step is coupling all the visualizations together using a dashboard which will enable the learner to understand the concept of the course. The tutor should take

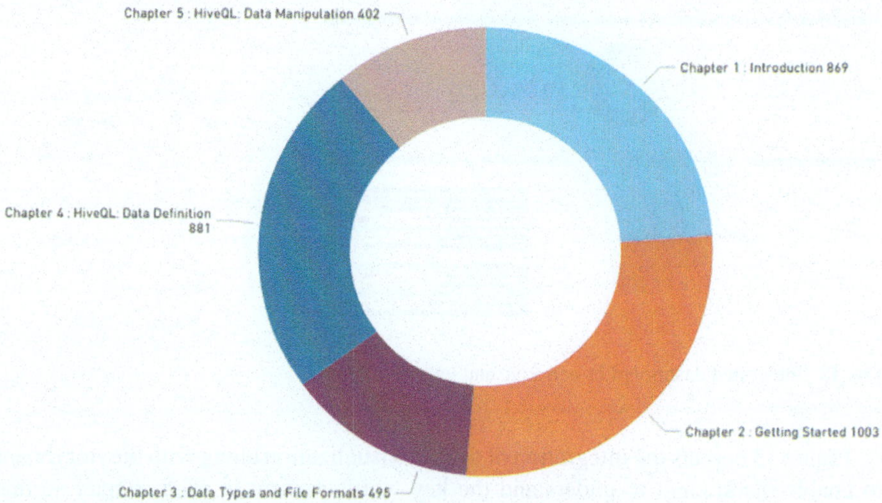

Fig. 11 Donut chart depicting the frequent words in each chapter

care that all the concepts available in the course material are provided to the student in a simple way. Figure 12 is the dashboard created through the storyboard of Canva. The tutor has taken care to explain the chapters by giving a title to the chapter which will enable the learner to have a glance on what is present in the chapter.

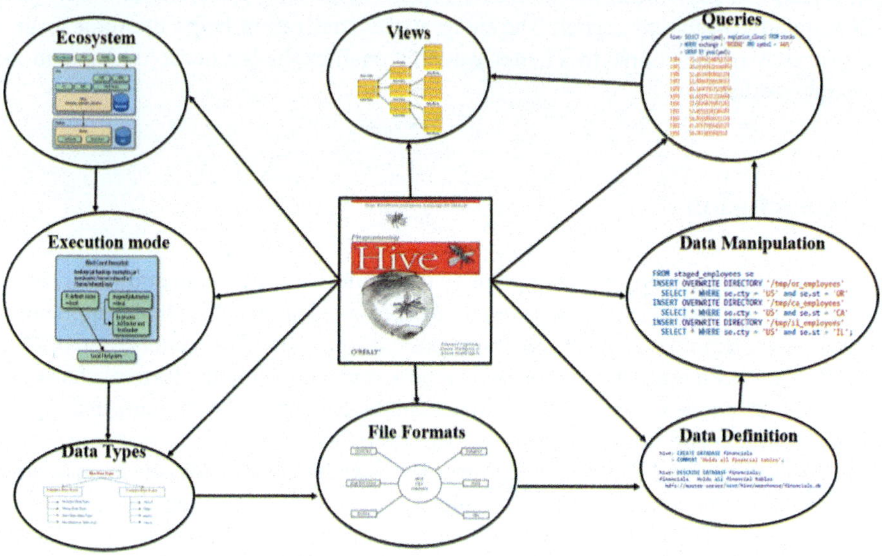

Fig. 12 Dashboard using Canva for the storyboard—course material

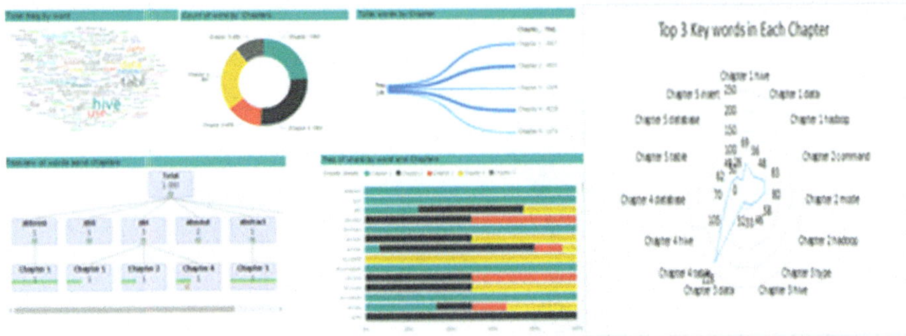

Fig. 13 Integrated dashboard of narrative and text visualization

Figure 13 depicts the integration of the text visualization along with the storyboard to enable the student to understand the key words present in each chapter so that student can understand the concepts available in each chapter. Therefore, it becomes the responsibility of the tutor to depict the appropriate keyword in the storyboard to explain the concept to the student. The word ecosystem gives an insight into the student that chapter one will be helping them to understand the Hive Ecosystem. Figure 13 depicts the integration of the dashboard integrating the narrative and text visualization of the course material.

In Fig. 13, the dashboard of narrative and text visualization has been coupled together to make the learner understand the concept easier. The wordcloud gives a list of frequent keywords present in each chapter. The other visualization charts have been already discussed earlier. The integrated visualization helps the tutor to give an overview of the course in a better way that enables the learner to understand the concepts better.

4 Conclusion

Data visualization is one of the imperative techniques in the current era which can be applied on various domains. This chapter has provided an insight into the application of data visualization using narrative and text in the field of education. Narrative visualization has been implemented to extract the text and depict in the form of charts and dashboard to enable the students to understand the content of the chapters in a sophisticated manner. The implementation of narrative visualization has been done using Canva storyboard and coupled with Power BI. Text visualization has been applied in two phases: processing the text to tokenize, removing the punctuation, stop words, numerals, etc. LDA algorithm was applied to generate topics based on each chapter, and frequent words were extracted from each chapter. Text processing and analysis are implemented using R tool. The visualization of the same is implemented using Power BI. Narrative and text visualization are coupled together using dashboards which has given a clear visual module of the course material to the learner. The learner is able to understand the content and flow of the course, important key words in each chapter and understand the course in a better way. In future, visualization techniques can be further implemented to other domains of education also.

References

1. https://web.cs.wpi.edu/~matt/courses/cs563/talks/datavis.html.
2. Johnson, M. (1993). *The moral imagination*. Chicago: University of Chicago Press.
3. MacIntyre, A. (1981). *After virtue*. Notre Dame, IN: University of Notre Dame Press.
4. Segel, Edward, & Heer, Jeffrey. (2010). Narrative visualization: Telling stories with data. *IEEE Transactions on Visualization and Computer Graphics, 16*(6), 1139–1148.
5. Kosara, R., & Mackinlay, J. (2013). Storytelling: The next step for visualization. *Computer, 46*(5), 44–50.
6. Lee, B., Riche, N. H., Isenberg, P., & Carpendale, S. (2015). More than telling a story: Transforming data into visually shared stories. *IEEE Computer Graphics and Applications, 35*(5), 84–90.
7. Kucher, K., & Kerren, A. (2015). Text visualization techniques: Taxonomy, visual survey, and community insights. In *2015 IEEE Pacific Visualization Symposium (PacificVis)*. ISSN: 978-1-4673-6879-7.
8. Kwon, B. C., Stoffel, F., Jäckle, D., Lee, B., & Keim, D. (2014). VisJockey: Enriching data stories through orchestrated interactive visualization. In *Poster Compendium of the Computation + Journalism Symposium* (Vol. 3).
9. Hullman, J., Diakopoulos, N., & Adar, E. (2013). Contextifier: Automatic generation of annotated stock visualizations. In *Proceedings of the SIGCHI Conference on Human Factors in Computing Systems* (pp. 2707–2716). ACM.
10. Bryan, C., Ma, K.-L., & Woodring, J. (2017). Temporal summary images: An approach to narrative visualization via interactive annotation generation and placement. *IEEE Transactions on Visualization and Computer Graphics, 23*(1), 511–520.
11. Metoyer, R., Zhi, Q., Janczuk, B., & Scheirer, W. (2018). Coupling story to visualization: Using textual analysis as a bridge between data and interpretation. IUI 2018, March 7–11, 2018, Tokyo, Japan. Copyright © 2018 ACM. ISBN: 978-1-4503-4945-1/18/03.
12. Pócsová, J., Mojžišová, A., & Takáč, M. (2016). Application of the visualization techniques in engineering education. In *17th International Carpathian Control Conference (ICCC)*, IEEE, INSPEC Accession Number: 16107928.
13. Teaching with Visualizations, Created by Bob MacKay, Clark College, https://serc.carleton.edu/sp/library/visualizations.
14. Klerkx, J., Verbert, K., & Duval, E., Enhancing learning with visualization techniques. Katholieke Universiteit Leuven, Belgium. https://core.ac.uk/download/pdf/34578672.pdf.
15. Baikadi, A., Goth, J., Mitchell, C. M., Ha, E. Y., Mott, B. W., & Lester, J. C., Towards a Computational Model of Narrative Visualization, AAAI Technical Report WS-11-18.
16. Segel, E., & Heer, J., Narrative visualization: Telling stories with data.
17. Alhawiti, K. M. (2014). Natural language processing and its use in education. *International Journal of Advanced Computer Science and Applications, 5*(12).
18. Litman, D. (2016). Natural language processing for enhancing teaching and learning. In *Proceedings of the Thirtieth AAAI Conference on Artificial Intelligence (AAAI-16)*.
19. Paul, M., & Girju, R. (2009) Cross-cultural analysis of blogs and forumswith mixed-collection topic models. In *Proceedings of the 2009 Conference on Empirical Methods in Natural Language Processing* (pp. 1408–1417), Singapore, 6–7 August 2009.

Data Visualization and Analysis for Air Quality Monitoring Using IBM Watson IoT Platform

K. S. Umadevi and D. Geraldine Bessie Amali

Abstract Data visualization is the common term that is used to describe presenting data in a pictorial or graphical format. Data visualization software helps to easily uncover patterns, trends and correlations that might go unnoticed in text-based data. It helps to break down the details of the data and make the data more comprehendible effortlessly. Another emerging field of study is Internet of Things (IoT). IoT deals with the difficulty of diverse device types, various sensors and types of data that gets generated and analysed in real time. An individual managing IoT has neither time nor the patience to decode the information at leisure. Unless the data is comprehendible, taking fast and result-oriented actions is difficult. That is the reason why data visualization becomes fundamental and viable. One of the tasks here is to be able to choose the form of visual depiction that works the best for the information. IBM's latest contribution to the field of data science brings analytics to your data and not the other way around. They build solutions that accumulate data from any type of source, including web and social. With those generated solutions, one can store, investigate and give an account of information by using analytic engines to drive actionable insights and visualization. In this work, we are making an attempt to communicate, understand and analyse data from societal application as well as enable big data analysis visualization to support real-time data.

Keywords IBM Watson · Cloud · Visualization

1 Big Data Analytics

The world is becoming progressively computerized day by day, so we literally manage, share and store our lives digitally. Online information is accumulated from our devices, computers and smart phones, and these devices are collecting and transmitting information on what we do and that is only the beginning. Nowadays, most of

K. S. Umadevi (✉) · D. Geraldine Bessie Amali
School of Computer Science and Engineering, Vellore, India
e-mail: umadeviks@vit.ac.in

D. Geraldine Bessie Amali
e-mail: geraldine.amali@vit.ac.in

S. M. Anouncia et al. (eds.), *Data Visualization*,
https://doi.org/10.1007/978-981-15-2282-6_2

the household gadgets including TVs, fridges, washing machine, watches and other devices are capable of collecting and transmitting data. In fact, if we accumulate every bit of the information from the earliest usage point until 2000, it would be not as much as what we presently make in a moment [1].

Big data may be precious for professional sake; it may even offer a new dimension for the success of professional development for the clients that were not recently envisioned. To process big data, we need huge number of computers and interconnected servers along with highly effective and efficient algorithms. These may help to evaluate a huge amount of data in few minutes. For example, Netflix analysed the information of their audiences for a better understanding of their popular shows and viewing patterns. Based on this information, they were able to create a fruitful series with the seamless blending of on-screen characters, directors and storyline. The traffic of big data is being examined to build up a vehicle that can drive totally accident-free independent from anyone else. It also helps to predict client behaviour, enhance and optimize business processes. But there are many difficulties:

- How to untangle the strands of enormous information and choose the significant parts of information which originates from such huge numbers of sources?
- How to realize where to visualize and access?
- How to transform data into knowledge?

To break down such vague, vast and bulk information, analytics on big data is accomplished through specialized software, tools, techniques and applications for prediction, mining and optimization. Though the procedures adopted are separate, they are highly integrated functions of high-performance analytics. Using the existing tools and software facilitates, an organization can process extremely large volumes of data gathered by the business people to figure out which information is important and needs to be investigated to drive better business choices later on.

1.1 Applications of Big Data Analytics

As demonstrated by Datamation, the recent developments in analysing big data enable analysts to decipher DNA of human in few minutes, foresee where terrorists intend to attack, figure out which gene is likely for specific illnesses and, obviously, which advertisements sound better to react to Facebook. Another precedent begins from one of the greatest mobile carriers on the planet. France's Orange launched its data for development venture by releasing subscriber data for clients in the Ivory Coast. The 2.5 billion records, which were made unidentified, included details on calls and texts conveyed between 5 million customers. Analysts got the information and sent Orange recommendations how the data could fill in as the establishment for the foundation of improvement venture to upgrade general prosperity and security. The proposed ventures included one that demonstrated to enhance public safety by following mobile phone information to track individuals after emergencies; another demonstrated to utilize cell information for disease containment.

1.2 Benefits of Big Data Analytics

Enterprises are dynamically planning to find important bits of knowledge from their data. Numerous big data projects begin from the need to answer explicit business questions. With the correct big data analysis, an enterprise can support deals, increment effectiveness and enhance activities and efficient client administration. QuinStreet surveyed 540 venture executives associated with big data purchases.

A major number of the respondents said that they were applying big data analytics to enhance client retention, assist them over their product development and gain a high ground over the competition. Specifically, 62% of respondents said that they use big data to improve speed and reduce complexity. Driven by specialized analytics software, big data can indicate the manner in which distinctive business benefits, including new openings for income, better benefits for clients, enhanced operational proficiency and gaining market trend over competitive rivals.

On a wide scale, data analytics technologies and techniques provide methods for breaking down information collected and drawing conclusions about them to enable enterprises to make educated business choices. Business intelligence inquiries answer essential questions concerning business operations and performances.

1.3 Challenges of Big Data Analytics

For most of the associations, big data analysis is a challenge. Consider the sheer volume and the various format of the data (structured as well as unstructured) gathered by the organization for various purpose. Now, this collected data needs to be consolidated, contrasted and examined to discover the various patterns and other profitable business data associated with it [2].

Big data analytics applications frequently incorporate information from internal as well as external sources, may be a weather-related data or a statistical data on consumers compiled by third-party information services providers. But, streaming analytics applications are becoming common in big data environments since it attracts more customers and reflects in real life. Users hope to do real-time analytics on data fed into Hadoop systems through Spark's Streaming module or other open-source stream processing engines, such as Flink and Storm.

Earlier big data systems were mostly installed on-premises, especially in large organizations that were gathering, organizing and analysing gigantic amounts of information. But cloud platform vendors, such as Amazon Web Services (AWS) and Microsoft, have made it easier to set up and manage Hadoop clusters in the cloud and have Hadoop suppliers such as Cloudera and Horton works, which support their distributions of the big data framework on the AWS and Microsoft Azure clouds. Clients would now be able to spin up clusters in the cloud, run them for as long as needed and then take them offline. It also does not require on-going software licences. Potential pitfalls comprise a lack or absence of internal analytics skills and

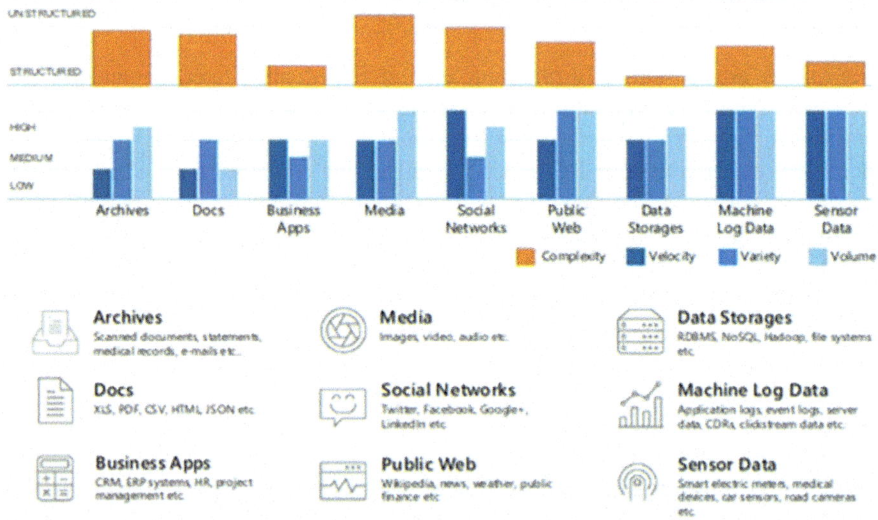

Fig. 1 Big data challenges

the mind-boggling expense for hiring experienced data scientists and data engineers to fill the gaps [3].

The amount of data that is typically involved, and its assortment, can cause data management issues in areas including data quality and consistency. Data silos can also result from the utilization of different platforms and data stores in big data architecture. What is more is integrating Hadoop, Spark and other big data tools into a cohesive architecture that meets an organization's big data analytics needs. It is a challenging proposition for many IT and analytics teams, which have to identify the correct blend of technologies and then put the pieces together (Fig. 1).

Problems in managing data and coordinating analytics efforts were among the top challenges to successful big data implementations cited by survey respondents (Fig. 2).

The rest of the chapter is organized as follows: Second chapter presents an overview of data visualization; next chapter elaborates the requirement of Air Quality Index. Last two chapters give information about the IBM Watson cloud, accessing the information and processing them.

2 Data Visualization

Data visualization is a representation of information or data in a graphical format [4]. The primary advantage of data visualization is not that it makes data looks more beautiful but it provides insight into complex data sets by communicating

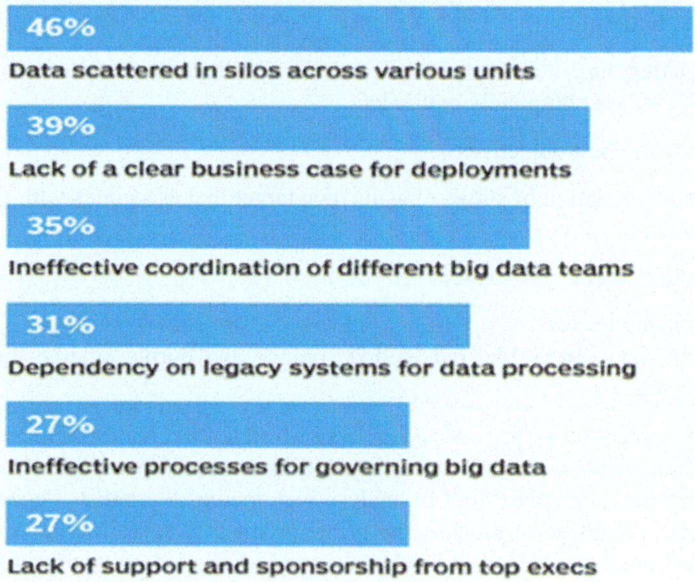

Fig. 2 Responses of 226 analytics, IT and business professionals

their significant aspects and more instinctive meaning in many ways. A simple form of data visualization includes graphs, pie charts and maps. To make a good data visualization, consider a series of numbers and count the number of 7 s within few seconds. Now imagine that we are not looking for specific numbers, but patterns. The power of design can be used to communicate information better.

Visual data makes it less demanding for the individuals to comprehend what a presenter is attempting to pass on. The various types of data visualization include charts like pie, line and bar, plots like scatter, whisker and stem, time lines, histograms, box and so on. Some of the simple steps that are followed to make an outstanding visualization are

i. Let it be straightforward:
   - Clearness is the most simple but most basic part of visualization
   - Flavouring adds some information but makes charts or graphs increasingly harder to decipher.

ii. Using appropriate tools
   - Make use of suitable business intelligence software
   - Use suitable software for the diversified audience.

iii. Narrate Story
   - Use text only when it is important to clear up something that is being visualized.

iv. Use of Ordering

- Relating the means of recorded and regarding them as increasingly/less critical; not completely equivalent.

v. Choose the best objectives

- Each perception or subject has its own target that associates with the central matter.

vi. Consider colour

- Engage viewers
- Stunning colours like red, yellow, orange and purple emerge the most in graphs.

Computer technology has assisted analysis of content to create graphical representations of noticeable concepts such as word, diagrams or infographics. Infographics have turned into a standard tool to analyse web media. Journalists depend incredibly on data visualization tools as the prime premise of revealing news about our general surroundings. Hence, it has been identified as a key to twenty-first-century research skill. Nowadays, anybody with a spreadsheet and an illustration application can make data visualization. There are also some extraordinary online tools that can help one to get started, e.g., many eyes is a free online data visualization tool created by IBM. Clients can choose from existing data collections or transfer their very own data and look over a visualization type. Many eyes breaks these into helpful classifications, for example comparing sets of values, rises and falls over time, see partly or in entirety. Gapminder is a free online service that gathers information and allows clients to choose world trends and compare them using a bubble chart. Gapminder is a very good tool to envision population trends. For advanced computer users who may be interested in high charts, Bigcharts is available. Bigcharts.com is an online data visualization service that is free for non-business use.

2.1 Process

Graphic designers can use visualization for perception, and analysts can convey their information more viably by understanding the visual structure standards behind the information. The techniques by themselves are not new, but rather their inference inside individual fields has kept them from being used together. We require a procedure that connects the individual discipline, attention and thoughts on how information is seen instead of the perspective in every individual field. The process of understanding information starts with a lot of numbers and inquiries. The associating steps shape them appropriate response:

- Acquire: Get the information, regardless of whether from a file on a stand-alone system or over a networked system.
- Parse: Give some structure to the information's significance and sort it out by their categories.
- Filter: Expel everything except the required information.
- Mine: Apply strategies from statistics or mining as a way to observe patterns or place the information with respect to numerical context.
- Represent: Pick a fundamental visual model, for example, bar graph, list or tree.
- Refine: Enhance the essential picture to make it clearer and add more information.
- Interact: Include strategies for controlling the information or obvious highlights.

3 Air Quality Index

Air is the consistent pressure that can be sensed but cannot be seen. It can be perceived through nature, by the undulating plants and trees and feeling the breeze. Though we cannot see the presence of air, its presence can be felt but considered as the most important for the survival of human beings [5]. Earlier, it was believed to be a single substance yet now it is realized that air is a blend of numerous gases. A blend of two prominent gases basically nitrogen and oxygen in a specific rate is the cardinal thing that enables us to live in such a maintainable condition on planet earth.

Air pollution is the contamination of air with detrimental gases, excess of particulate matter and other substances. It leads to diseases and allergies for humans and animals and poses a threat to the natural habitat. Air pollution arises due to man made as well as natural processes. The man-made sources of air pollution include power plants that run on fossil fuels, motor vehicles, burning of garbage, slash burn in forests and decomposition of waste in landfills (generates methane). Natural sources would include dust (because of lack of vegetation), radioactive decay, wildfires, etc. The most common pollutants include sulphur dioxide, carbon monoxide, nitrogen oxides, carbon monoxides, particulate matter (PM_{10}, $PM_{2.5}$), chlorofluorocarbons, etc. According to the World Health Organization (WHO), air pollution levels rise hazardously high in many parts of the world, and a recent data from WHO declares that 9 out of 10 people inhale air having high levels of impurities [6]. Pollution in urban cities causes epidemic and increases at a high rate. Also, WHO assessment depicts that nearly 7 million individuals pass away every year due to polluted air. Ambient air pollution alone caused 4.2 million deaths in 2016. Household air pollution caused by cooking with polluting fuels caused an estimated of 3.8 million deaths.

Air quality is among the top priorities to be tackled because people are facing several issues due to air pollution. The air is polluted by the pollutants like PM_{10}, $PM_{2.5}$, sulphur dioxide and nitrogen dioxide. Having polluted by such a large number of pollutants, air is increasing the risk of respiratory diseases among citizens. World Health Organization (WHO) reported that India is one among the highly polluted countries, since 13 of the most polluted cities around the world are located in India including New Delhi. Various respiratory diseases like asthma, bronchitis and sometimes lung cancer can curtail from living in cities that have severe air pollution. Cardiovascular diseases can also be caused by the presence of air pollutants including cardiac arrest. Many premature birth and death are known to be rooted from air pollution. A study revealed that air pollution is ranked as fifth in causing more number of deaths in India [7]. The number of diseases directly linked to air pollution is also considerably high. The central pollution board has announced stricter norms for pollutants.

An Air Quality Index (AQI) is a number forecasting the pollution rate to the public by the government agencies. The increase in the pollution rate indicates the higher health risk for the public. Hence, various countries use difference index to represent the concern air quality. For example, Canada, Malaysia, Singapore and India use Air Quality Health Index, Air Pollution Index and the Pollutant Standards Index and the National Air Quality Index, respectively. Air quality is the number one concern for the residents of New Delhi. It has been named the most polluted city in the world with its Air Quality Index (AQI) being double of the upper limit of the hazardous level on the scale. The pollution particles cause smog in the city that has resulted in numerous road accidents. Breathing in the city is equivalent to smoking 50 cigarettes a day, in terms of lung damage. The air pollution data of New Delhi provided by the Central Pollution Control Board to observe its trends and changes with respect to other meteorological factors like relative humidity, ambient temperature, wind direction, wind speed and solar radiation may be analysed. This will be helpful for the experts to understand some of the pollutants like nitric oxide, suspended particulate matter, nitrogen dioxide, ozone, sulphur dioxide (SO_2), $PM_{2.5}$, toluene, benzene, para-xylene, ortho-xylene, oxides of nitrogen (NO_x), ammonia (NH_3) and their impact.

In New Delhi, the Air Quality Index has gone up to 999 which are considered to be higher than the upper limit in negative scale. The American embassy has installed a more sensitive sensor (air quality instrument), and it was able to measure a reading of the AQI as high as 1010. During November 2017, the US airlines cancelled all its flights connecting to New Delhi due to the poor air quality. Medical practitioners warned that breathing highly polluted air is like breathing 50 cigarettes per day. The harmful gases and excess amounts of particulate matter in the air have choked the residents of the city which comprised up to 19 million citizens. The situation has been declared as a public health emergency as the lives of a lot of people are at stake. A 20% increase in the number of patients, has been diseases caused by air pollution, has been observed. In the year 2015, about 9 million deaths were reported with cause rooted at air pollution. And among these, over 2.5 million happened in India. This is an alarming figure and shows that this is a medical emergency.

The growth of urbanization and fast increase in development of New Delhi has had adverse effects on the quality of air in the city. A lot of studies and researches have been performed in wake of the situation. S. Taneja and N. Sharma found patterns in air pollution using linear regression and multilayer perceptron network [8]. They were able to find patterns among the various different toxins like sulphur dioxide, nitrogen dioxide, PM_{10}, $PM_{2.5}$ and carbon monoxide. The studies showed that the concentration of PM_{10} will increase by 45.9% in the future.

R. Bhardwaj and D. Pruthi performed time series and predictability analysis of the pollutants in the air of New Delhi [9]. They have estimated the regression coefficient, Hurst, fractal dimension and predictability index of various different pollutants. The Hurst will be helpful in finding out whether it is regressing strongly to the mean of the data or to the cluster and recognized using fractal measurement. Fractal measurement gives an idea about the harshness of a surface. The results depict that carbon monoxide follows Brownian movement with sulphur dioxide and relative humidity. This behaviour is prevalent to the before and after odd-even scheme which was conducted by the government of New Delhi to curb air pollution. Because of this nature, it was concluded that it is difficult to predict a pattern of the various pollutants.

Saksena et al. have performed cluster analysis on air quality data of New Delhi which was data spanned over a period of nine years on which they performed spatial classification [10]. Classification was done for three prevalent pollutants observed—PM, nitrogen dioxide and sulphur dioxide. The algorithm preferred was a hierarchical agglomerative along with Euclidean distance and concluded that the two classes were prevalent. They observed that the average concentration of these pollutants is usually the same irrespective of where the station is located and encouraged finding the systematic bias in the data collected.

Anikendar Kumar and Pramita Goyal attempted to forecast the levels of pollution in New Delhi using principal component analysis using data spanned over a period of seven years [11]. They measured the air quality at ITO in New Delhi, which is a very important junction in the city and usually has huge amount of standing traffic. They calculated the covariance of the matrix using the data input. Eigen values were then checked, and only those components which had them to be calculated as greater than 1 were used further for the prediction of AQI. The prediction was done by principal component regression method. Their study showed that using this method of prediction, the values were more accurate during the winter season than the rest. The normalized mean square error, which is used to measure accuracy, was 0.0058 for winter season, 0.0082 for summer, 0.0241 after monsoon season and 0.0418 during monsoon.

M. Statheropoulosa and N. Vassiliadisa analysed air quality data spanning a period of five years [12]. One of the air monitoring stations in Athens is where the data was recorded. The data included concentrations of various pollutants including carbon monoxide, nitric oxide, nitrogen dioxide, ozone, particulate matter, sulphur dioxide, etc. The data also included values of many other meteorological components. They then performed principal component analysis on this data to find out the features that were more impactful. They found that these main features turned out to be correlated with combustion of oil and gasoline and the use of ozone. They performed a canonical correlation analysis on the data. They also had an inclination air dryness and fast winds. Most importantly, it was found out that the pollution was related highly with humidity and speed of wind.

Jerrett et al. performed spatial analysis of air pollution on mortality of the city of Los Angeles [13]. They assessed the exposure of pollutants in the air by keeping a track on the common ones. They attempted to measure the effects on health caused by bad air quality. They got data from the American Cancer society on 22,905 subjects over a time span of 1982–2000. They performed calculation on the spatial multilevel Cox regression models. They concluded that there was a 95% relative risk of mortality with increase of 10 $\mu g/m^3$ $PM_{2.5}$. So, an increase in $PM_{2.5}$ and other particulate matter among other pollutants can influence mortality rate quite significantly.

4 IBM Watson IoT Platform: Turn Numbers into Narratives

What makes handling data so difficult and complex is that, without anyone else's input, do remain meaningless. Data is inactive; it is not self-arranging or even self-comprehensive. Data is the base with minimal measure of apparently helpful. Data is more valuable than information, knowledge is more valuable than data, and wisdom is noteworthy estimation above all. Data requires something unique—a program, a machine or even an individual—to add value and become more informative. Visualizing data is indispensable. The more prominent quantity and varieties of information gathered, the more we have to try different things to make it unique. Instead of starting from the beginning, start from a clear page and explore different avenues regarding a custom visualization.

Business intelligence tools convince that the perfect procedure to make representations is to load information in a device, pick a random chart out of the tool box and complete the task in few clicks. However, simplified solutions can be framed for issues that are hard-to-characterize so as to solve them. User attitude reveals us that choices are not founded by any one or single information. Rather, we will in general look at and integrate numerous kinds of information before arriving a conclusion. Due to our inclination towards analysis, clarity does not come at the same time; whereas, we frequently search to determine meaning.

Data driven does not mean the fact that collected information and the tools used to gather it are human made. So, they are not in pure form, yet the channels used are very subjective. Hence, we should have enough capacity to digest the reality and help us comprehend it, as indicated by various factors they may be unique or consistently changing. A visualization technique is not proposed to be highly contrasting, yet sufficient for exploration, since it conveys importance to the general meaning to the data as opposed to getting hung up on the numbers.

- ***Characterize your expectation with supporting clients***: Recognize your supporting clients and become more acquainted with their issues, so as to convey precise messages, attempt to comprehend the subject extremely even becoming a domain expert whenever possible. At that point, identify user's expected models, the information they manage and account the task that needs to be finished. Depend on basic graphs in advance and do not be threatened by Excel sheets with huge measures of information. Characterize a point of view or principle as per your objective and reason.

- ***Clean and Understand***: Begin to have a glimpse at your informational collection for its structure and the typologies of information included (strings, numbers and dates). Examine different rows and columns by watching their relationship and implications. Clarify the requirement and how your information will be incorporated inside the tool or product. Ask yourself: Is the information going to be continuous? Will there be any periodic update? Will future updates have any variations and which "ranges" would it be a good idea for me to consider? To clean information, examine the irregularities, clean the duplicates and check for character encoding. Use tools for data checking and validate to proceed with refinement of your data set. Give careful consideration to likeness between elements in your data set.
- ***Model and check for visual validity***: Check the underlying hypothesis: Group the data set using appropriate data structure suitable for your application. Identify the various attributes associated with or nature of data and observe its behaviour. Try to apply various analysis techniques that may be primary, secondary or any more meaningful manipulations to remodel the structure of the data. Identify the pattern using the simple visual representation in order to verify the formulated hypothesis. The simplest models are preferable when the data set is too huge and willing to identify the patterns. It also helps to focus more on the real topic and entry point of interest.
- ***Structure and style***: The heart of visualization lies in the decisions made by the user about where to emphasize and hide details so as to effectively convey the message. Avoid maintaining strategies from preconceived notion, automation and alternate solution instead focus on the content. Try to start from scratch, identify new version for the design and variety of charts to present the data set. Presentation of data is collaborative practices that will enable to structure in order to drive valuable experiences. Consider the advancement in the procedure adopted and try not to restrict to basic chart models. Next, work with shading palettes, typography and line graphs to make your representation look like IBM and be agreeable. Consider the key standards while examining:

  - *Expressiveness guideline*: Say all that you need to state—no more, no less—and do not delude.
  - *Viability rule*: Use or make the best technique accessible to demonstrate your information.

- ***Test and repeat***: Set up heuristic assessments and usability tests. Walk through the models according to client's requirement and run the model for small and large data set for simple cases. Collect the observation and opinion from supporting clients, who are not related to the project. Summarize the view and incorporate the changes required in the visuals.
- ***Refine and implement***: Search for bugs and functional mistakes from the visuals. Work together with your design team in order to decide the best tools and libraries for building the information vis. Once survey the structures in code for availability, globalization and so on.

5 Visualizing Using IBM Watson IoT Platform

Visualizing information in graphical ways can give you insights into your data. By enabling you to look at and explore data from different perspectives, visualizations can help you identify patterns, connections and relationships within that data as well as understand large amounts of information very quickly (Fig. 3).

- Column charts are utilized to depict correlations among things or information changes over a period of time. Data represented using column or rows can be represented using a column chart in a sheet.
- Bar charts are utilized to illustrate connections among individual things. Data represented using column or rows can be represented using a bar chart in a sheet.
- Pie chart can be utilized to show the relationship of parts to the aggregate. Data represented using column or rows can be represented using a pie chart in a sheet.
- Area charts are utilized to emphasize the degree of change over time. One can likewise utilize area charts to demonstrate the relationship of parts to a whole. Data represented using column or rows can be represented using area chart in a sheet.
- Line charts can be utilized to relate consistent data over a period of time at a typical scale. Consequently, line charts are perfect for exhibiting information patterns at

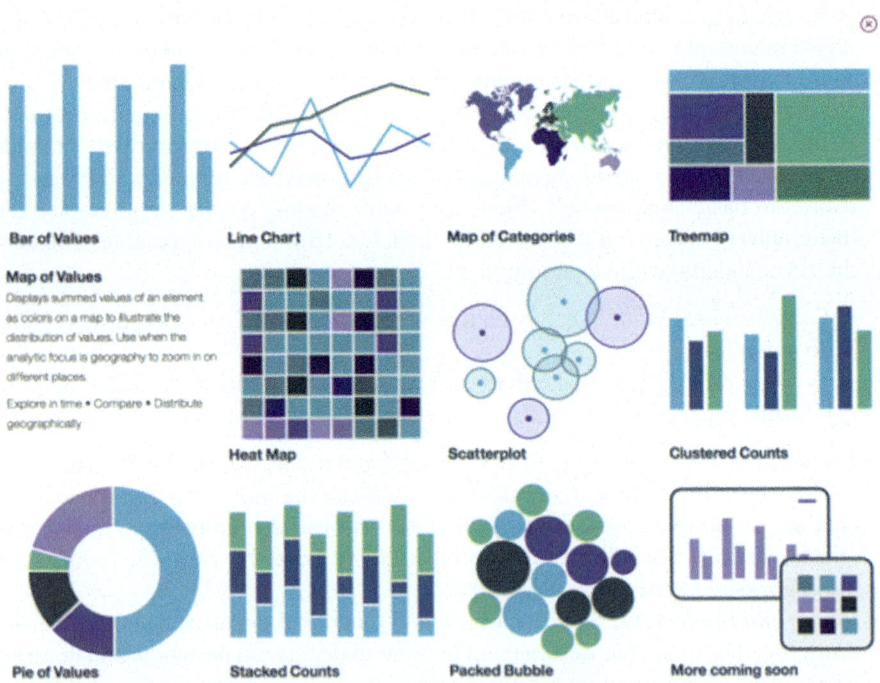

Fig. 3 Data visualization supported by IBM cloud

equivalent interims. Data represented using column or rows can be represented using line chart in a sheet.

- Scatter charts show the connections among the numeric values in a few information arrangements or show two groups of data as a single series of XY coordinates. Data represented using column or rows can be represented using scatter chart in a sheet.
- Box plot charts compare distributions between many groups or data sets. They display the variation in groups of data: the spread and skew of that data as well as outliers.
- Candlestick charts are a type of financial chart that displays price movements of a security, derivative or currency.
- Customized charts give you the ability to render charts based on JSON input.
- Dual Y-axes charts use two Y-axis variables to show relationships between data.
- Error bars indicate the error or uncertainty in a value, and they give a general idea of how precise a value is or conversely, how far a value might be from the true value.
- Heat map charts display data as colour to convey activity levels or density. Typically, lower density values are displayed as cooler colours; whereas, higher density values are displayed as warmer colours.
- Histogram charts show the frequency distribution of data.
- Map charts show geographic point data, enabling you to compare values and show categories across geographical regions.
- Multi-series charts display data from multiple data sets or multiple columns as a series of points connected by straight lines or bars.
- Parallel coordinate charts display and compare rows of data (called profiles) to find similarities. Each row is a line, and the value in each column of the row is represented by a point on that line.
- Population pyramid charts show the frequency distribution of a variable across categories. They are typically used to show changes in demographic data.
- Quantile–quantile (Q-Q) plot charts compare the expected distribution values with the observed values by plotting their quantiles.
- Relationship charts show how columns of data relate to one another and what the strength of that relationship is by using varying types of lines.
- Scatter plot charts show correlation (how much one variable is affected by another) by displaying and comparing the values in two columns.
- Scatterplot matrix charts are scatter plot charts organized into a matrix so that it is easy to look at all pairwise correlations together.
- t-SNE charts enable you to picture high-dimensional data sets. They are useful for embedding high-dimensional data into a space of a few dimensions, which can then be visualized in a scatter plot.
- Treemap charts display hierarchical data as a set of nested areas. Use to compare sizes between groups and single elements nested in the groups.
- Word cloud charts display how frequently words shown in content by making the extent of each word corresponding to its frequency occurrence.

Case Study:
The air pollution data will be analysed to observe its trends and changes with respect to other meteorological factors like relative humidity, ambient temperature, wind direction, wind speed and solar radiation. Some of the pollutants observed are nitric oxide (NO), suspended particulate matter (PM_{10}), nitrogen dioxide (NO_2), ozone, sulphur dioxide (SO_2), $PM_{2.5}$, toluene, benzene, M&P xylene, oxylene and ammonia (NH_3). The following steps need to be carried out to sense and visualize the data.

Step 1: Configure the IoT device to the IBM Watson IoT Platform Cloud environment.
Step 2: Create board and submit (Fig. 4).
Step 3: Click on the pollution data board and add new card (Fig. 5).
Step 4: Select the mode of visualization (Fig. 6).

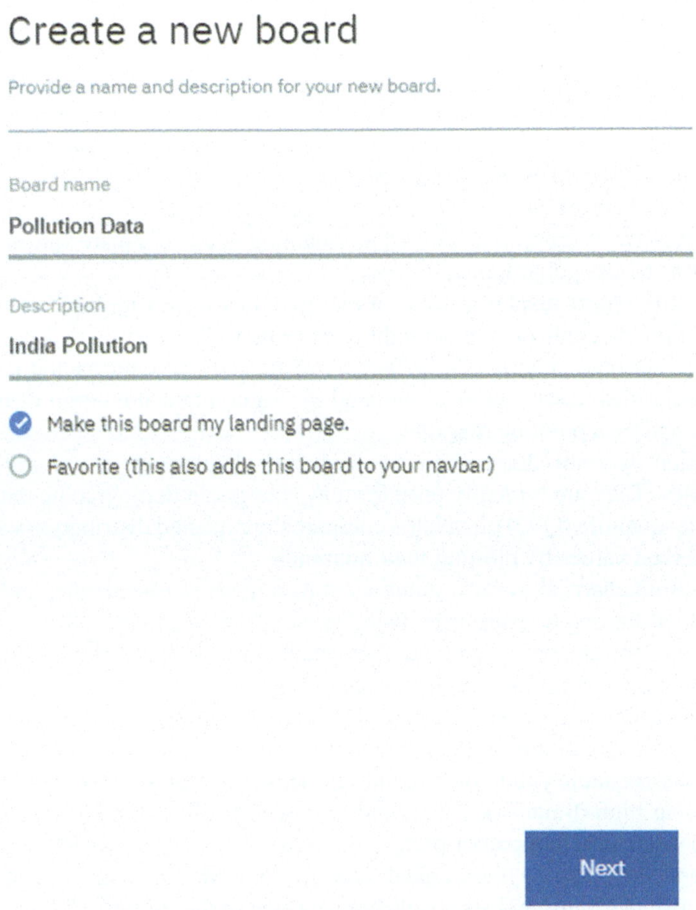

Fig. 4 Create board and click submit

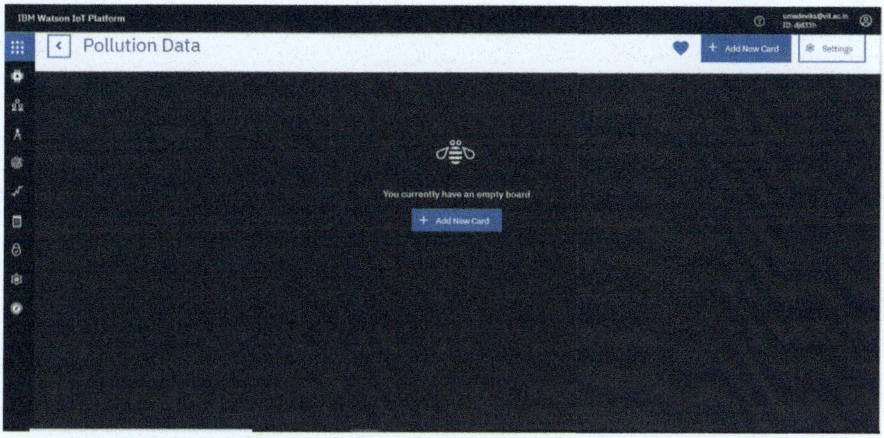

Fig. 5 Click on the pollution data board and add new card

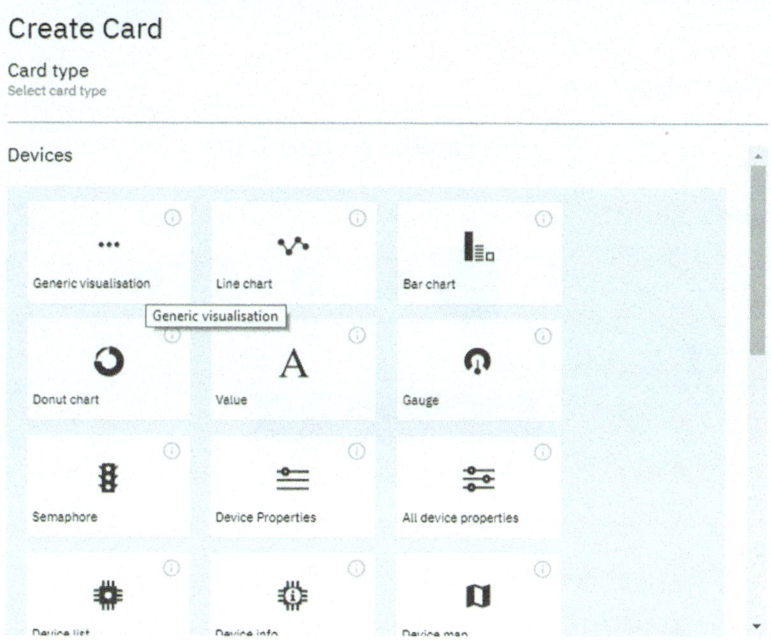

Fig. 6 Select the mode of visualization

Step 5: Select the input source, e.g., NodeMCU device used for sensing the carbon dioxide values (Fig. 7).
Step 6: Select the chart type, e.g., line chart and size as XL (Fig. 8).
Step 7: Choose the colour (Fig. 9).
Step 8: Observe the results (Fig. 10).

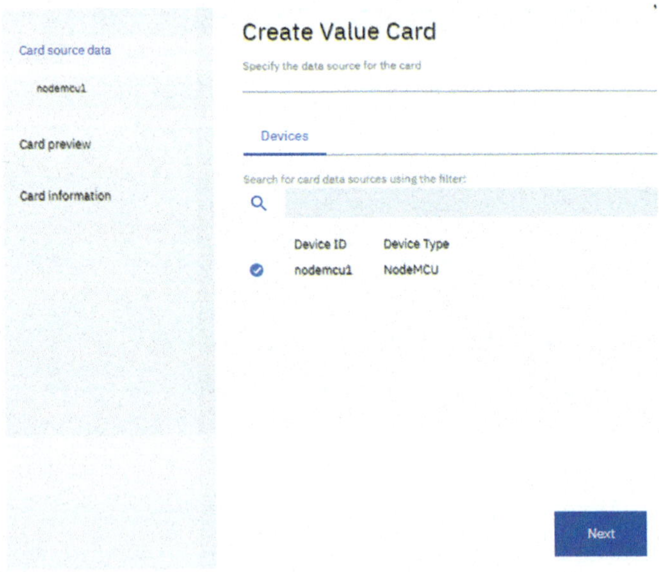

Fig. 7 Select the input source

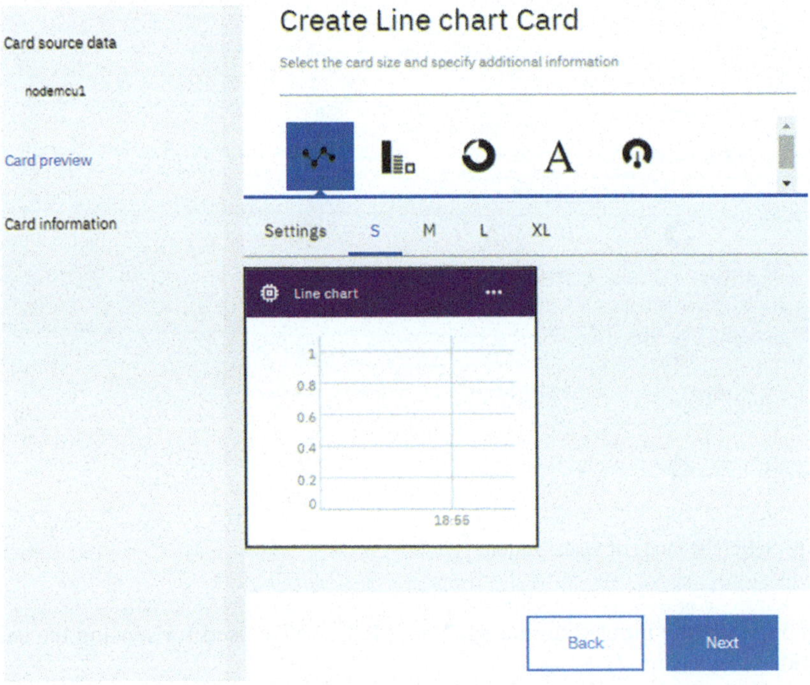

Fig. 8 Select the chart type

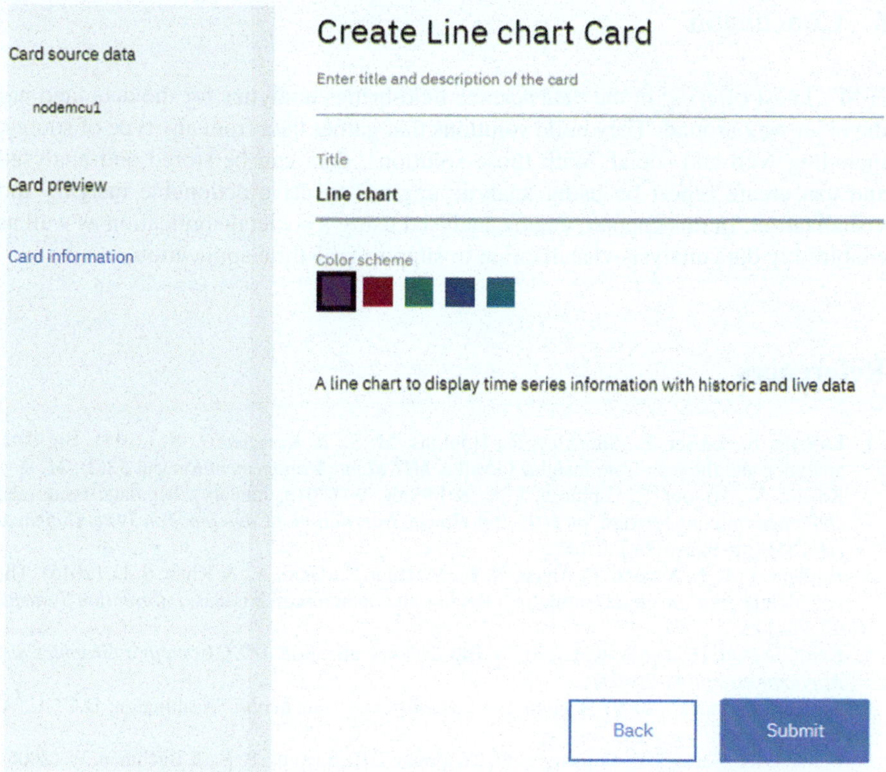

Fig. 9 Choose the colour

Fig. 10 Results

6 Conclusion

IBM's latest offering in the data science field brings analytics for the data, and not the other way around. They build solutions that gather data from any type of source, including web and social. With those solutions, data can be stored and analysed and can create report by using analytic engines to drive actionable insights and visualization. In this chapter, data is analysed using a societal application as well as enables big data analysis visualization to support real-time application.

References

1. LaValle, S., Lesser, E., Shockley, R., Hopkins, M. S., & Kruschwitz, N. (2011). Big data, analytics and the path from insights to value. *MIT Sloan Management Review, 52*(2), 21.
2. Kaisler, S., Armour, F., Espinosa, J. A., & Money, W. (2013, January). Big data: Issues and challenges moving forward. In: *2013 46th Hawaii International Conference on System Sciences (HICSS)* (pp. 995–1004). IEEE.
3. Hashem, I. A. T., Yaqoob, I., Anuar, N. B., Mokhtar, S., Gani, A., & Khan, S. U. (2015). The rise of "big data" on cloud computing: Review and open research issues. *Information Systems, 47,* 98–115.
4. Keim, D., Qu, H., & Ma, K. L. (2013). Big-data visualization. *IEEE Computer Graphics and Applications, 33*(4), 20–21.
5. Air Quality Index. (2009). *A guide to air quality and your health*. Washington, D.C., USA: EPA.
6. Frank, L. D., Sallis, J. F., Conway, T. L., Chapman, J. E., Saelens, B. E., & Bachman, W. (2006). Many pathways from land use to health: Associations between neighborhood walkability and active transportation, body mass index, and air quality. *Journal of the American Planning Association, 72*(1), 75–87.
7. Gadani, H., & Vyas, A. (2011). Anesthetic gases and global warming: Potentials, prevention and future of anesthesia. *Anesthesia, Essays and Researches, 5*(1), 5.
8. Taneja, S., Sharma, N., Oberoi, K., & Navoria, Y. (2016, August). Predicting trends in air pollution in Delhi using data mining. In: *2016 1st India International Conference on Information Processing* (IICIP) (pp. 1–6). IEEE.
9. Bhardwaj, R., & Pruthi, D. (2016, October). Time series and predictability analysis of air pollutants in Delhi. In: *2016 2nd International Conference on Next Generation Computing Technologies (NGCT)* (pp. 553–560). IEEE.
10. Saksena, S., Joshi, V., & Patil, R. S. (2003). Cluster analysis of Delhi's ambient air quality data. *Journal of Environmental Monitoring, 5*(3), 491–499.
11. Kumar, A., & Goyal, P. (2011). Forecasting of air quality in Delhi using principal component regression technique. *Atmospheric Pollution Research, 2*(4), 436–444.
12. Statheropoulos, M., Vassiliadis, N., & Pappa, A. (1998). Principal component and canonical correlation analysis for examining air pollution and meteorological data. *Atmospheric Environment, 32*(6), 1087–1095.
13. Jerrett, M., Burnett, R. T., Ma, R., Pope III, C. A., Krewski, D., Newbold, K. B. … Thun, M. J. (2005). Spatial analysis of air pollution and mortality in Los Angeles. *Epidemiology*, 727–736.

Comparative Analysis of Tools for Big Data Visualization and Challenges

G. Divya Zion and B. K. Tripathy

Abstract The equivalent term for data visualization is 'visual communication,' representing data visually. The theme of data visualization is to convey information clearly, efficiently to users by using graphics, and it helps to have an inner view of data. Data visualization mesmerizes users by changing tedious data into a visually colorful tale. For this, we need data visualization tools that are popular in understanding data easily and visually. When analyzing data, data visualization is one of the steps that aroused to present the data to the users. Big data has come up considerably in recent times and has become an integral part of modern database research. In fact, there is a need for analyzing large amount of data which is temporal. Other characteristics of these datasets are that these are dynamic, noisy and heterogeneous in nature. As a result, to transform these different types of datasets into an accessible and understandable format, we need big data visualization tools. Traditionally, in the beginning, visualization in big data area was done by traditional systems but this approach was not capable of handling variety in datasets. Also, these traditional systems are restricted to deal with the datasets of small size. So, here comes the need for modern systems to analyze variety of datasets in big data, and this modern system gives awareness of modern data visualization tools that supports a variety of big datasets. Henceforth, the comparative analysis on visualization tools and challenges allows user to go with the best visualization tool for analyzing the big data based on the nature of the dataset. Since their inception, several tools have been proposed to model and analyze the process of data visualization in big datasets. In this chapter, we propose to analyze these tools in the form of their strengths and weaknesses. Also, we have planned to discuss in detail their applications and suitability in dealing with different situations (Caldarola, Picariello, & Rinaldi in Experiences in wordnet visualization with labeled graph database. Berlin: Springer, pp. 80–99, [1], Caldarola, Picariello, Rinaldi, & Sacco in Exploration and visualization of big graphs—The DBpedia case study. KDIR (IC3K 2016), pp. 257–264, [2], Caldarola, & Rinaldi in

G. Divya Zion
School of Computer Science and Engineering, VIT, Vellore, Tamil Nadu 632014, India
e-mail: gdivya.zion2016@vitstudent.ac.in

B. K. Tripathy (✉)
School of Information Technology and Engineering, VIT, Vellore, Tamil Nadu 632014, India
e-mail: tripathybk@vit.ac.in

S. M. Anouncia et al. (eds.), *Data Visualization*,
https://doi.org/10.1007/978-981-15-2282-6_3

Improving the visualization of wordnet large lexical database through semantic tag clouds. IEEE, pp. 34–41 [3]).

Keywords Big data · Data visualization · Different tools for data visualization

1 Introduction

Visualization is a familiarized term which is meant for representing data in the form of graphics using pictorialism which is pictorial of design. Visualization goal is where information is made easy to understand. Data visualization provides a facility where patterns with complex structure could be shown or explained using different shapes, lines and different colors. In the year 1987, the foundation of National Science at the USA has started their research in visualization, and these researchers have defined visualization as, "The analysis of the measured scientific data is represented using Computer Graphics." In the year 1989, the Oxford English Dictionary stated for the term visualization as, "anything that can be seen through the mind sense of ability, an image or a picture which is either visible to the eyesight or which is visible to our mind or our imagination is called visualization [4]."

Visualization is meant for three purposes

(1) Exploring

To explore data which is unknown or it is called as data exploration.

(2) Analyzing

Analyzing helps us to show proof or which is used either for falsification or verification.

(3) Presenting

To present everything that is known about the data or presenting output.

1.1 Data Visualization

It can be defined as representing raw data, unprocessed data and information using graphical elements such as charts, data visualization tools, graphs and maps. As human beings, we quickly draw our attention to designs and colors by differentiating one color with that of the other and one shape with that of another shape. We got habituated to a culture of visualizing beginning with art, advertisements in movies and TV. Therefore, data visualization clutches to the interest of users by representing "data" as "visual" by using visualization tools [5]. The way of representing the data which is normal into understandable and useful data can be called as data visualization. In case if the data size exists in thousands of varieties, then it becomes difficult to implement meaningful data. Below is an example, where a vegetable capsicum has different variations of colors and is very difficult to find the variations of colors this capsicum represents to data of various formats like age, income, address

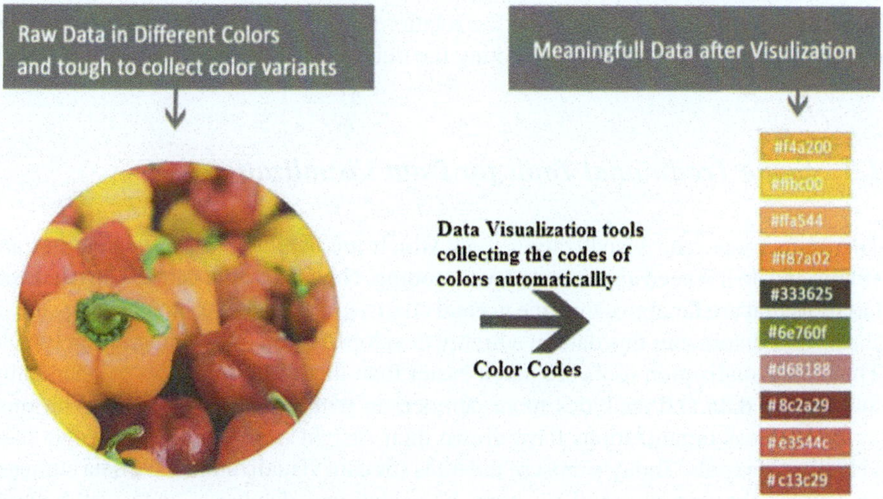

Fig. 1 Process of data visualization

and time produced from different sources which is not in an understandable format. So, it is impossible to read, analyze and visualize the data that is collected; hence, data visualization is more recommended to bring out data in an understandable format (Fig. 1).

1.2 The Need for Data Visualization

Usually organizations and industries generate tremendous amounts of data by taking support of web search engines, and this leads to dramatic increase of data availability in the Internet. As the size of data increases, users feel difficult to analyze, understand and make use of the available data. The solution for this is to make the data to be visualized so that users can understand the data. So, visualizing the data can be considered as "visual communication" where data is represented visually [6]. The tools for data visualization are necessary for performing analytics, as these tools are easier to work than that of the traditional systems. So, organizations develop their own data visualization tools for their understanding. The softwares and tools that are used for data visualization are used for big data and its analytics, as the data in big data needs a quick way to understand and has an overview on the data. So, obviously visualization tools have become an easy way to do it. When visualization is performed on the data, we need to ensure that the visualization tools are performing as they are meant for [7–9].

The data visualization tools are of many types, and they report the output visually by interpreting the results. The visualization tools absorb the data from different sources and put graphic on the data. Some visualization tools explain the shape of the data automatically and find out the relationship among those variables and then discover the type of chart for presenting the output [10].

1.3 Some Traditional Tools for Data Visualization

There are some data visualization tools which are available as softwares that are simple and do not need any knowledge on coding. Nowadays, the data to be visualized has tools that are far above than the normal charts, graphs that are shown using spread sheets and displaying this data in a highly developed manner by using infographics. The data visualization tools are much easier than the traditional tools to operate by making the data and analytics more democratic within an organization. With this, many business organizations have drawn their insight into building their own data visualization tools. Today's most of the tools for data visualization have data storages or sources that are attached to them, like relational databases, data warehouses, Hadoop and various cloud storages. This data visualization tools software takes the data in from different data storages and applies graphic type on the data, this software provides a way for the user to represent the data, and these tools automatically explain the data shape that is used. The below are some visualization software tools like Tableau, Power BI and many more which have excellent performance and allow the software do your task. These tools use JavaScript as their programming language and are available at lower cost.

The below are some visualization software tools that are easy to use and allow the software do your task.

1. **iCharts**

iCharts provides a way for industry and market to meet consumer experts. It presents the output in the form of charts like Google docs data, spreadsheets for the people of business, sports and many more categories. This tool makes use of cloud that allows many companies to put their data in the cloud and view it as charts to many viewers through the web. Users can access the iCharts for free trial, and paid version is also available which has many more features provided in it [10].

2. **Fusion Charts Suit XT**

Fusion Charts Suite XT is a higher end tool for data visualization that uses library files like JavaScript chart that provides facility to create charts of any type. It has more than 90 types of charts, and these different types of charts satisfy individual requirements of users like changing font, border, color and scrolling, clicking, zooming and taking print. These charts can be displayed in 2 and 3 dimensions.

3. **Modest Maps**

Modest Maps is a tool available in a package and provides different functionalities like maps that are interactive and free libraries for the developers who make their maps to look interactive in their projects.

4. **Raw**

This data visualization tool is a free software web application tool. It acts as a mediator between spreadsheets and vector graphics. This tool considers a dataset and checks for hierarchy, it supports different charts, and the installation of this tool is pretty simple which makes everything work from client's perspective.

5. **Tableau**

It is a data visualization tool that is used for business purposes; this tool is fast and flexible. This supports all types of data formats and connects to different servers. Different types of charts are available, and the user interface is very flexible. It does not need any coding techniques but simply it allows running the input in R, and the outputs are imported to Tableau. This importing part needs some basics on programming skills based on the given input.

6. **Microsoft Power BI**

This tool is a user-friendly that is used for the purpose of business analytics by taking the help of cloud. The services that are provided by this tool are flexible to the users so visualization can be done within minutes. This tool makes use of different tools like MS Office, SQL and SharePoint. The advantage of this tool is it allows user to put queries to the data using natural language which does not need any programming language and providing input through R script is optional. The data from multiple sources is combined and are easy to use.

7. **Gephi**

Gephi is a user-interactive data visualization tool and is applicable for different networks and complicated systems. It produces the output graphs either dynamic or rank based [2]. This tool is capable of handling large datasets and produces good visualization outputs in the form of graphs by sorting the data into a meaningful and understandable format. This tool does not need any programming skills but basics in graphics are a must.

8. **Plotly**

This tool uses Python and can be called as Plotly. It is used for analyzing and makes the data to visualize. Users can access this tool for free but are allowed to have only some features. Dashboards and charts are created online and can be accessed through offline. As every tool uses charts to represent its output, similarly it also provides facility where output can be generated using different charts. This Plotly tool uses another tool for automatic intake of data from the static image.

9. **Excel 2016**

As we know Excel is a spreadsheet that is used for statistical analysis in big data, it is able to handle data which is semi-structured format. The graphs that are present in Excel make the tool as user-friendly.

10. **HighChartjs**

This tool uses library files for representing the outputs in charts using JavaScripts. This works in all browsers which are new, and this does not require programming knowledge but uses a pair of keys and their values kinked with colons and separated by commas.

1.4 The Weaknesses of These Tools

The performance of these tools can be considered as good only when some resources of that tool are available, and this leads to incorrect output. The weakness of some tools is explained below:

iCharts, Fusion Charts Suit XT, Modest Maps and Raw are simplest and user-friendly tools, and all of them represent the output in charts based on their prescribed format. The weakness of Tableau is though it is available for free the services are available with storage limitation up to 1 GB. To have all the services of the tool to be available, user needs to pay for it and access it. And to have visualization analysis for large datasets, coding or in-depth programming knowledge is necessary.

Microsoft Power BI, as we know it is used for business purpose using cloud services; for this we need to buy the cloud where public cloud for free service will not be allowed to do analysis. The performance of this tool is slow when compared to that of Tableau and when accessing the cloud services they are limited up to 250 MB.

Plotly, for free versions it provides less uploading capacity like up to 500 KB, for paid users there are no restrictions where users can access charts unlimitedly to produce the outputs. To access this tool, users need to have programming knowledge.

Gephi tool is specialized in generating the output in the form of Graphs; this does not have a capacity to generate the output using different ways, and Excel 2016 is not a free tool to be accessed by the users [11].

2 Visualizing Big Data

2.1 Brief Introduction to Big Data

The term big data does not tell what exactly the size of database should be in order to say that the data is big. But, big data allows the new tools, technology to handle the data that is present in it within a period of time [12]. This term "big data" resembles or talks about datasets which are huge or difficult for processing using traditional systems. When the size of data increases, i.e., having more attributes (columns), there is a high chance of discovering data falsely. The traditional systems like databases, relational databases and data management systems have difficulty in handling data with huge volumes. To handle this all the servers need to work parallelly.

Big data contains datasets of size that are above the ability of a software tool to manage, process and capture the data within a stipulated time. Big data includes unstructured, structured and semi-structured data. In structured data, the data is processed, stored and retrieved in a fixed format, and the information is organized readily and can be accessed from a database using algorithms of search engine. For example, in a company database, employee table is in a fixed format that gives the details of an employee. Unstructured data refers to the data which lacks specific format or structure of data. This leads to time taking process in analyzing the data, e.g., email. Semi-structured data has data in both the formats, where data is not classified in any formats. Big data focuses mainly on unstructured data. Big data is seen in almost every field like finance, medical, banking, online shopping sites, business, data clustering [13–19], Social Internet of Things [20], Neighbourhood Systems [21], Social networks [22], Deep learning [23] and life sciences [24–27].

Haenlein, definition for big data, the huge amounts of datasets and the continuously updated different formats of data (numeric's/texts/images) are characterized based on the characteristics of big data (Fig. 2).

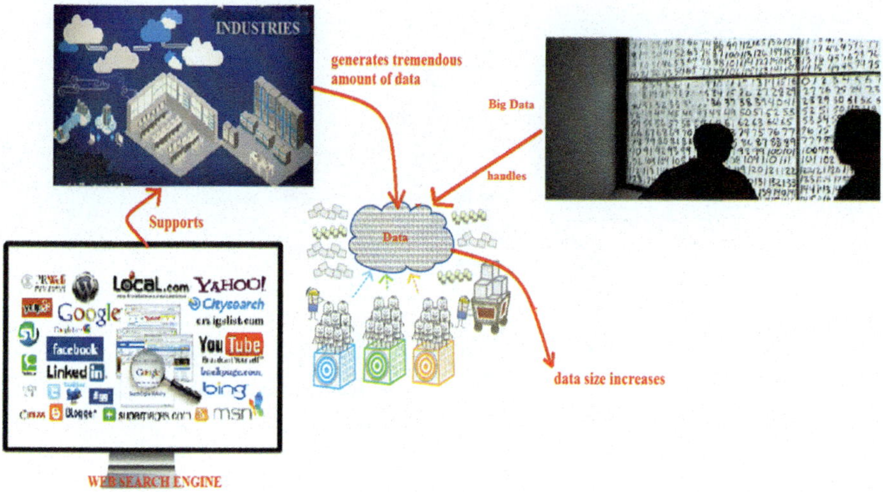

Fig. 2 Process of generating data and handled by big data

2.2 Characteristics of Big Data

The specialty of big data is its characteristics. There are many characteristics that describe big data but the below listed are the most essential ones.

(1) Volume: It focuses on the quantity of data that is being generated and storing it. This quantity of data ascertains the value and capacity for a data to be considered as big data.
(2) Variety: It focuses on types/contents/formats of data. The data like text, audio, video and images is also considered as big data.
(3) Velocity: It focuses on the speed at which the data is generated and collected. In big data, the generation of data is continuous, in terms of frequency of data generation and handling.
(4) Veracity: It focuses on the quality of data in such a way that the data that is captured differs greatly by affecting the accuracy in analysis.
(5) Value: It focuses on useful data, from the available data.

So, all the characteristics of big data in real time are handled by tools for analytical purpose and as the data is not in an understandable format we impose visualization tools on the big data. Below is an image which shows the characteristics of big data (Fig. 3).

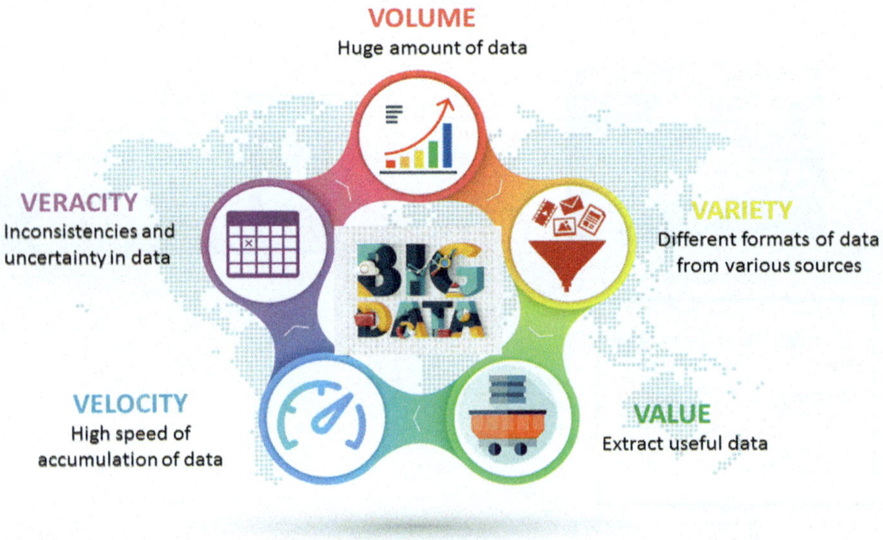

Fig. 3 Characteristics of big data

2.3 *Handling Large Data Volumes*

The technical components that handle big data from the point of Gartner convey that big data increases day to day based on it high volume, high velocity, variety and veracity. All the characteristics of big data are scalable, parallel and cost effective. There were times where data need to be extracted from databases and loaded into data warehouses using traditional methods. But, this procedure falls behind when working with big data [28].

When the huge data is continuously generated, big data takes the help of Hadoop File system for storing the data as data warehouses and database management systems are not able to handle it. Hadoop File System processes the data either by Hadoop else by Spark, where both are considered as analytical engines. To handle the large data volumes of big data, there are some technologies that are emerging which are explained below [29].

(1) To process Big Data,
    MapReduce and Hadoop frameworks are considered for processing and storing big data.
(2) To perform analysis, querying on data,
    WibiData, Pig is considered.
(3) For business intelligence,
    Hive is considered which uses SQL to perform analytics in Hadoop File System.
(4) For machine learning,
    Apache Mahout and SkyTree both are considered as machine learning algorithms which use HDFS.

(1) MapReduce:

It is a framework used by Hadoop, which is used to divide the work among the group of similar nodes. It has two tasks Map and Reduce tasks; Map considers a set or group of data into another group of data so that each element is split into number of individual tuples. In Reduce, the output of maps is considered as input and combining these group of tuples into small group of tuples. So, we can conclude Reduce task is done after the Map task. The advantage of this framework is that data can be processed for many number of nodes (Fig. 4).

(2) Hadoop:

Hadoop is an Apache framework, which is used for processing large datasets that are distributed among group of similar nodes. Hadoop is used to work using a single server connected to thousands of systems where each system provides storage and computation. Hadoop has two processing layers, MapReduce and Hadoop Distributed File System. In Hadoop, data is partitioned into files, directories of size 128–64 M. These files are then sent to group of similar nodes for processing (Fig. 5).

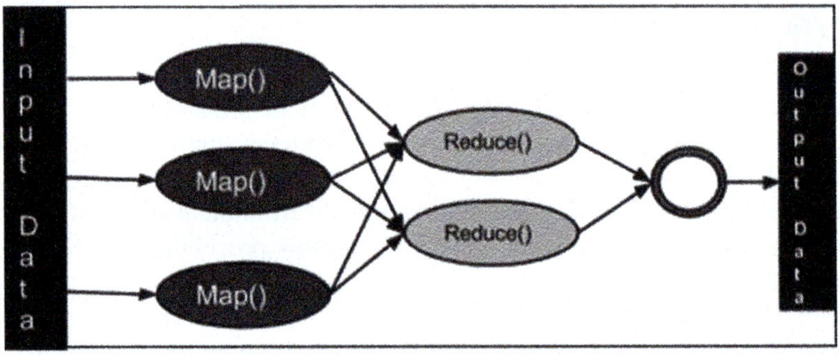

Fig. 4 MapReduce framework

(3) Pig:

Pig is a technology offered by Apache for the purpose of MapReduce parallel programming jobs, and they are executed on nodes that are similar to Hadoop. This provides a platform to examine datasets of large size. It provides user-defined functions for processing, and queries are posed on the data that is stored in Hadoop. The language which Pig uses us Pig Latin, which uses Java programming for doing MapReduce tasks (Fig. 6).

(4) Hive:

Apache Hive is a software that provides SQL interface to post a query on the data which is stored in databases and files. This enables business intelligence applications to put queries on group of similar nodes in Hadoop (Fig. 7).

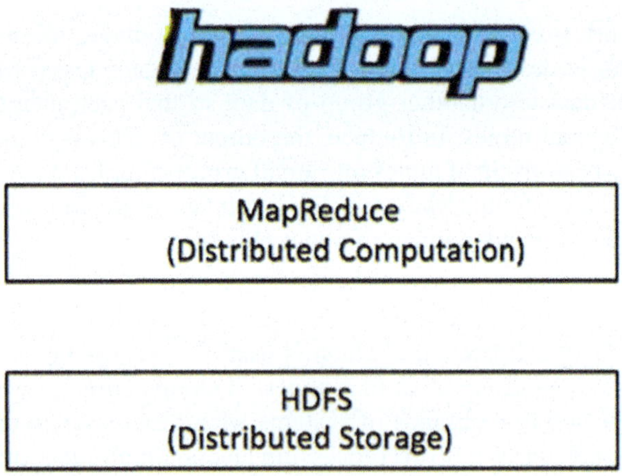

Fig. 5 Hadoop framework

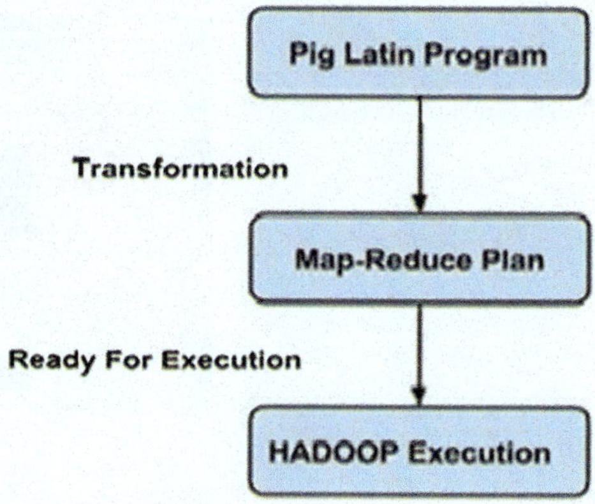

Fig. 6 Working of Pig

Fig. 7 Working of Hive

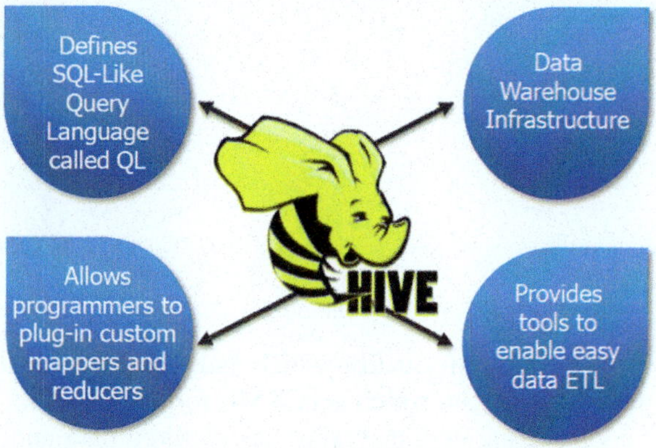

2.4 *Visualizing Semi-structured and Unstructured Data*

Unstructured Data:

Unstructured data focuses on the data which does not have any specific format or structure, and this takes time to analyze the data as it is unstructured. Unstructured data is usually a heavy data which has dates and numbers [30]. Examples of unstructured data are

(1) Blogs, emails and social media
(2) Web logs and reviews of any product
(3) Documents that are in text (Fig. 8).

Semi-structured Data:

Semi-structured data contains data that is present in both structured and unstructured data. In this though the data is not divided under a database for storage yet it contains

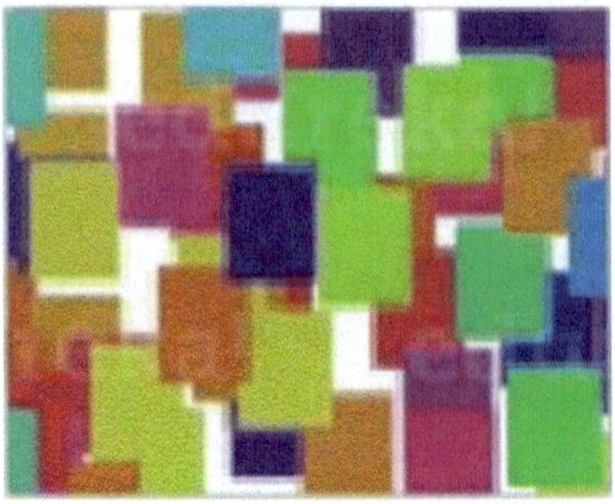

Fig. 8 Unstructured data

Fig. 9 Semi-structured data

the important information which makes individual attributes to segregate within the data e.g., the data which is in XML format (Fig. 9).

Visualizing and analyzing unstructured data and semi-structured data are done using search analytics. In search analytics, there are many technologies like search index, ElasticSearch and Apache Solr which access both structured and unstructured data. Search analytics is the one that allows us to filter from large datasets. Based on the search algorithm, the search will be accurate and uses search engines to improve the outputs. Zoomdata is a tool which is built for the unstructured data in big data along with the tools like Hadoop and Cassandra that are supported by big data [30].

Big Data is a combination of both semi-structured and unstructured data. ElasticSearch is the process which does analysis of such type of datasets. It provides results which are similar to the output of a query. It uses NoSQL which has a role similar to that of SQL for relational databases. Searches for an answer using NoSQL is performed as soon as the data is uploaded.

3 Visualization Tools for Big Data

As huge amount of data is generated, there are many datasets available at big data and the nature of datasets is dynamic, heterogeneous and noisy. So in order to transform these different types of datasets into an accessible, understandable format is a difficult process. Hence, we move forth to data visualization which is known as visualization in big data. The visualization in big data is done in traditional and modern systems, and the pros and cons of both the systems are discussed below [31]. The reason why we as users do not understand data is, as much of the raw data is available and the size of this raw data will be created in the coming years. With this huge amount of raw data, it is difficult to get an analytical solution, so analytics through big data provides a solution. Even then if humans are not up to the mark to explain the data which is interpreted using analytics, then the analytics tools that are considered as best go into vain. The data visualization tools of big data are being a necessity in many fields nowadays.

There are data visualization tools, for users to make their data to be visualized within minutes. Now the job of users is to select the right one (tool) that can be best suited for their (user) needs. The output can mesmerize users by the tool they have used and can be able to handle large datasets. But the jackpot is everything depends on the right tool that suits your project.

Some of the positive insights that make users to access visualization through big data [32]:

(1) Processes huge data quickly
(2) Makes the data to be understood and responding to the user
(3) Narrates story from the data.

3.1 Drawbacks of the Traditional Tools

Traditional systems,

(1) Not able to handle the size of many datasets.
(2) These systems are restricted to deal with datasets which are smaller in size.
(3) These are limited to access the preprocessed sets of static data and are operated only in offline mode.

Modern systems,

(1) Able to handle big dynamic datasets, with limited resources like storage and computational.
(2) Visualizing datasets which are larger in size is a task in modern systems.
(3) Modern systems should satisfy many preferences of users, requirements and tasks posted by the users.
(4) This systems should get automatically adjusted to the parameters of the environment settings, screen resolution and the size of memory that is available.

3.2 Tools Available for Data Visualization in the Context of Big Data

The users of data visualization tools and analyst of data should have the below things (Table 1).

There are some tools that are discussed in Sect. 2, and all those tools present their output in the form of charts or they are desktop-oriented applications. Here are some data visualization tools, and these tools provide users an interface to interact with the computing environments like R and MATLAB.

(1) Commetrix:

It is a framework that is considered for visualization in networks, and analysis is done by researchers of network area. This tool gets data from all sources of network like business networks and has electronic communication and discussions using voice over IP, emailing.

(2) SocNetV:

It is a social networks visualizer tool, which is user-friendly in order to perform analysis on social networks. This allows user to build networks on a virtual platform and allows networks of various formats.

(3) Sentinel Visualizer:

This tool is mainly used for geospatial mapping, and this allows user to check many layers of links among entities and model relationships among them. It provides features for drawing entities and highlights them.

Table 1 Data experts features

	Description
Core skills	• **Analytical**: strong mathematics and statistics foundation
	• **Technical**: programming and database skills
Domain and business knowledge	• **Knowledge of the sector**: understanding data sources and real-world situations and processes behind the data
	• **Awareness of business goals and processes**: knowing the business questions that matter, and how data fits with the organization
Soft skills	• **Storytelling**: ability to transform analytical insights into actionable—and compelling—business recommendations
	• **Team working**: enjoy working with people from different disciplines
Competencies	• **Analytical mindset**: being able to re-formulate complex questions as analytical tasks
	• **Creativity**: knack for generating unexpected solutions to problems and exploring data from different angles
	• **Curiosity**: wanting to understand how the world works

(4) Tulip:

It is a framework which is used for analyzing and visualizing relational data. It provides libraries for interactive visualization. The framework is written in C++ language that allows the development of new algorithms, data models, visual encodings and particular domain visualization techniques.

(5) Visone:

This is the software used for creating visuals, representing networks data. The aim of the software is to make researchers get enhanced in social technology and then perform analysis and visualize the networks data in a different manner.

Many more data visualization tools are available at market, based on user preferences of input data; user can select the visualization tool.

3.3 Technical Competencies of These Tools

The essential tools and technical competencies are not defined in an exact manner. It depends on organization to have a specific domain and a certain prescribed tools based on the requirement. When performing analytics, the tools which produce user-interactive results will be considered more. The requirement of tool is based on situation and the type data, and this way of doing data visualization can be considered as best format for handling big data. Gathering multiple brains into visual feedback is the best way of importing the knowledge; when the size of data increases beyond the limit, then setting boundaries in order to intake that particular data will be actionably changed.

The top most skills that are necessary in data visualization are as follows:

(1) The people who are called as experts should need to have broad range of programming skills.
(2) Some basic analytical skills are required in order to have solutions for your queries.
(3) Sorting the answer for a correct question.

With the use of data visualization tools, user comes across using multiple technologies like cloud storages, sharing resources like tools and extensions facilities, and all of them are imposed on the data of various formats. The output produced by any of the tool looks like the outputs produced by the other tools.

The difference between traditional tools and the modern visualizations tools is cost. The modern tools provide many facilities for the user only when licensing, and cost for entire product or tool is paid. The traditional tools do not have this lope pole but user needs to pay for development part which is not done in modern tools. However, the traditional tools do not have warranty, which make users to fall at risk. Both the traditional systems and modern systems have pros and cons, but everything depends on the right tool that suits your project.

4 Challenges

4.1 Gaps in Research in Finding Tools for Visualization of Big Data

Visualization is a wide area which does not have an exit as there are very swift advances. Visualization and its analytics are a proposition that is supposed for a human to exist and improves the outcome and the quality of results significantly. The term big data is a hot buzz word where we can see everywhere in today's technology in magazines, tweets, articles and blogs. As the size of data increases and the data being generated from different sources, there is a possibility of appearing different datasets that are large in size and are difficult to handle using traditional system which can be overcome by big data [33].

As the data volume that is available is large, we opt for big data which presents new chance for advancement in analysis and new levels of thinking. There is an increasing significance in scientific research on visualization tools for big data that is developing rapidly [34]. A few gaps in research that are identified are

(1) An increase in use of big data, so there is a need to understand and evaluate that data.
(2) An act of increasing big data databases and big data analytical tools in order to classify the mapping tools.
(3) The effect of measuring the big data analytical tools by the researchers and publishers, so that the evaluation report explains the strength of analytical tool based on the researchers feedback.
(4) The evaluation report of researchers plays a social impact in the research such that the technology rate increases based on its performance toward the users.
(5) As there are large datasets available that are needed to be analyzed by combining different types of datasets, there are many research questions that are needed to be answered by using these datasets.

Research on applications of Visualization in big data is in its early stage. So, it warrants more rigorous research in this direction and development of new tools for support.

Some of the steps to be taken in this direction are:

(1) Training humans on cognitive skills so that users can easily understand the visualized data.
(2) Conducting sessions for users with data experts on how to handle the visualization tool design errors and interpreting them to the end users.
(3) Develop specific visualization tools that are included as a part with command line interfaces so that the data experts allow users to have free interaction with the tools like zooming and clicking.

The data visualization can be done in two ways like static and dynamic visualization, where static visualization is preferred during starting stages of analytical procedure but the dynamic visualization is considered as a good one as the data is changed dynamically so the representation should be done dynamically. In such a way, the gaps between the visualization tools applying on big data can be resolved.

5 Future Scope

As we know, huge amount of data is being generated every minute and every year by industries, web and many more other organizations. So here comes the challenge to store, utilize and evaluate this huge data which are used in many business aspects. Storing the huge amount of data that is generated does not yield positive result, henceforth tools are needed for analyzing data and to handle the big data that is generated every day [35, 36].

Some challenges of big data visualization are as the data is generated from different sources, definitely there are datasets of different type and different size which are present [37–42]. Hence, there is a challenge to **synchronize datasets from different data sources**. Secondly, though the analytical tools provide us a way to understand the huge data using analytics, even then the **organizations need people with skills in order to explore the analysis of big data** to industries and organizations. Thirdly, as discussed earlier, **storing huge data is an everlasting challenge** that leads to privacy concerns, usually when the size of data grows data warehouses comes into picture that handles data with large quantities in structured and unstructured formats. This type of data leads to missing values and duplicates. Fourthly, there is a challenge to the developers of big data analytics that the tools that are being developed with good performance rate should be able to provide a convenient way for users to access data as there is uncertainty in data, and these tools should not lead users to risks and new problems with the uncertain data. There is no ending process for research to come out with a solution for these problems, so user can choose a best big data analytical tool according to the data, and there is a need to develop analytical tools that will be able to overcome all these challenges.

6 Conclusion

Analysis of various tools for big data visualization brings us a justification stating that when we look back to past years data was available, but not in high amount or quantity where spontaneous choice taken by humans was considered. When the data available is at high amount with respect to quantity, the rate of taking spontaneous choice was decreased; this state provided data not to disappear but rather keeps growing every minute and every day. The volume of data in traditional systems like data warehouses sometimes increases up to terabytes, and these huge volumes of data reduce the performance of data warehouse by consuming all of its resources. Based on this disadvantage of the traditional systems performance, big data technology has evolved, where big data is able to handle huge data but users feel difficult to analyze, understand and make use of the available data. So data visualization tools provide services like technical support and software for big data. The data visualization technology directs us to examine tremendous amount of data generated by organizations and industries by considering the subsequent past, evaluate present and predicting the future to have better and lucrative results. Sometimes, organizations choose software for data visualization but are not able to understand the outcome and henceforth end up in selecting different tools; this leads researchers to problems in comparing the different data and differences in previous outputs. Therefore, many data visualization tools are developed for users, business analysts and executives to analyze, understand and make use of the available data.

References

1. Caldarola, E. G., Picariello, A., & Rinaldi, A. M. (2015). Experiences in wordnet visualization with labeled graph database. In *International Joint Conference on Knowledge Discovery, Knowledge Engineering, and Knowledge Management* (pp. 80–99). Berlin: Springer.
2. Caldarola, E. G., Picariello, A., Rinaldi, A. M., & Sacco, M. (2016). Exploration and visualization of big graphs—The DBpedia case study. In *Proceedings of the 8th International Joint Conference on Knowledge Discovery, Knowledge Engineering and Knowledge Management* (Vol. 1, pp. 257–264), KDIR (IC3K 2016).
3. Caldarola, E. G., & Rinaldi, A. M. (2016). Improving the visualization of wordnet large lexical database through semantic tag clouds. In *2016 IEEE International Congress on Big Data (BigDataCongress)* (pp. 34–41). IEEE.
4. Chen, G., A short introduction on data visualization.
5. Wang, L., Wang, G., & Alexander, C. A. (2015). Big data and visualization: Methods, challenges and technology progress. *Digital Technologies, 1*(1), 33–38. Science and Education Publishing.
6. Shadare, A. E., Sadiku, M. N. O., Musa, S. M., & Akujuobi, C. M., Data visualization. *International Journal of Engineering Research And Advanced Technology (IJERAT), 02*(12). ISSN: 2454-6135.
7. Dadzie, A., & Pietriga, E. (2017). Visualization of linked data—Reprise. *Semantic Web, 8*(1).
8. Marie, N., & Gandon, F. L. (2014). Survey of linked data based exploration systems. In *International Workshop on Intelligent Exploration of Semantic Data.*
9. Dadzie, A., & Rowe, M. (2011). Approaches to visualizing linked data: A survey. *SemanticWeb, 2*(2).
10. https://www.fastcompany.com/3029239/30-simple-tools-for-data-visualization.
11. Ali, S. M., Gupta, N., Nayak, G. K., & Lenka, R. K. (2016). Big data visualization: Tools and challenges. 978-1-5090-5256-1/16/$31.00_c 2016 IEEE.
12. https://opensource.com/resources/big-data.
13. Tripathy, B. K., Vishwakarma, H. R., & Kothari, D. P. (2014). Neighbourhood based knowledge acquisition using MapReduce from big data over cloud computing. *Proceedings of CSIBIG 14*, 183–188.
14. Tripathy, B. K., & Deepthi, P. H. (2015). *Handling Fuzziness in Big Data using Clustering Techniques, NCICT-15, Bangalore.*
15. Tripathy, B. K., & Mittal, D. (2016). Hadoop based uncertain possibilistic kernelized c-means algorithms for image segmentation and a comparative analysis. *Applied Soft Computing, 46*, 886–923.
16. Tripathy, B. K., Deepthi, P. H., & Mittal, D. (2016). Hadoop with intuitionistic fuzzy C-means for clustering in big data. *Advances in Intelligent Systems and Computing, 438*, 599–610.

17. Tripathy, B. K., & Deepthi, P. H. (2017). An investigation of fuzzy techniques in clustering of big data. In V. Sugumaran, S. Arun Kumar & T. Arun Kumar (Eds.), *Computational Intelligence Applications in Business Intelligence and Big Data Analytics*. CRC Press, (Taylor & Francis Group).
18. Tripathy, B. K., Seetha, H., & Murthy, M. K. (2017). Uncertainty based clustering algorithms for large data sets. In *Modern Technologies for Big Data Classification and Clustering* (pp. 1–33). IGI Publications.
19. Seetha, H., Tripathy, B. K., & Murthy M. K. (2017). *Modern Technologies for Big Data Classification and Clustering*. IGI Publications.
20. Tripathy, B. K., & Dutta, D. (2018). Trustworthiness in the social internet of things (SIoT). In M. Panda, A. E. Hassanien & A. Abraham (Eds.), *Edited Volume- Big Data Analytics: A social Network Approach* (p. 18). Taylor and Francis Publisher.
21. Tripathy, B. K. (2017). Rough set and neighbourhood systems in big data analysis. In V. Sugumaran, S. Arun Kumar & T. Arun Kumar (Eds.), *Computational Intelligence Applications in Business Intelligence and Big Data Analytics*. CRC Press, (Taylor & Francis Group).
22. Tripathy, B. K., Sooraj, T. R., & Mohanty, R. K. (2018). Data mining techniques in big data for social network. In M. Panda, A. E. Hassanien & A. Abraham (Eds.), *Edited Volume-Big Data Analytics: A Social Network Approach* (p. 21). Taylor and Francis Publisher.
23. Adate, A., & Tripathy, B. K. (2018). Deep learning techniques for image processing. In S. Bhattacharyya, H. Bhaumik, A. Mukherjee & S. De (Eds.), *Machine Learning for Big Data Analysis* (pp. 69–90). Berlin, Boston: De Gruyter. https://doi.org/10.1515/9783110551433-003.
24. Weinberg, B. D., Davis, L., & Berger, P. D. (2013). Perspectives on big data. *Journal of Marketing Analytics, 1*(4), 187–201.
25. Caldarola, E. G., Picariello, & Castelluccia, D. (2015). Modern enterprises in the bubble: Why big data matters. *ACM SIGSOFT Software Engineering Notes, 40*(1), 1–4.
26. Caldarola, E. G., & Rinaldi, A. M. (2015). Big data: A survey—The new paradigms, methodologies and tools. In *Proceedings of 4th International Conference on Data Management Technologies and Applications* (pp. 362–370).
27. Caldarola, E. G., Sacco, M., & Terkaj, W. (2014). Bigdata: The current wave front of the tsunami. *ACS Applied Computer Science, 10*(4), 7–18.
28. Su, X., Introduction to big data. Institutt for informatikk og e-læring ved NTNU Learning.
29. Shukla, S., Kukade, V., & Mujawar, S. (2015). Big data: Concept, handling and challenges: An overview. *International Journal of Computer Applications (0975–8887), 114*(11).
30. https://www.zoomdata.com/product/unstructured-data-analytics-visualization/.
31. Bikakis, N. (2018). Big data visualization tools. ATHENA Research Center, Greece, arXiv: 1801.08336v2[cs.DB], February 22, 2018.
32. https://www.salesforce.com/hub/analytics/why-use-big-data-visualization/.
33. http://strata.oreilly.com/2010/01/roger-magoulas-on-big-data.html.
34. Research Trends, Special Issue on Big Data, Issue 30, September 2012.
35. https://elearningindustry.com/big-data-analytics-challenges-faced-business-enterprises-7-top.
36. https://www.scnsoft.com/blog/big-data-challenges-and-their-solutions.
37. Caldarola, E. G., & Rinaldi, A. M. (2017). Big data visualization tools: A survey—The new paradigms, methodologies and tools for large data sets visualization. Conference Paper, July 2017.
38. Data visualization techniques from basics to big data with SAS® Visual Analytics. White Paper.
39. Big data visualization: Turning big data into big insights. The Rise of Visualization-based Data Discovery Tools, White Paper, MARCH 2013, Intel IT Center.
40. Keim, D., Qu, H., & Ma, K.-L. (2013). Big-data visualization: Published by the IEEE Computer Society 0272-1716/13/$31.00 © 2013 IEEE.
41. Kumar, O., & Goyal, A. (2016). Visualization: A novel approach for big data analytics. In *2016 Second International Conference on Computational Intelligence & Communication Technology*.

42. Jena, B. (2017). A review on data visualization tools used for big data. *International Research Journal of Engineering and Technology (IRJET), 04*(01). ISSN: 2395-0056.
43. Liu, S., Maljovec, D., Wang, B., Bremer, P., & Pascucci, V. (2017). Visualizing high dimensional data: Advances in the past decade. *IEEE Transactions on Visualization and Computer Graphics (TVCG), 23*(3).

Data Visualization Techniques: Traditional Data to Big Data

Parul Gandhi and Jyoti Pruthi

Abstract Data visualization is one of the interactive ways that lead to new innovation and discovery. It is a dynamic tool that opens new ways of research which facilitate the scientific process. With extensive use of the Internet and Web, a large amount of data is generated every day. There is a need to understand large and complex data. When the data is available in large volume, it has to be processed by using various data processing methods and need to present it with different types of techniques and methods. Data visualization is a key to the success of any enterprise as it helps enterprises to control the data in an effective manner and make the best utilization of that data to convert it into knowledge. It is a process of converting data and numbers into visual form. Data visualization techniques use different effects of computer graphics. It helps the stake holders to make an effective and fast decision making. It also provides the better understanding for pattern recognition, analysis of trends, and to extract the appropriate information from the visuals. Visualizing data may be a challenge but it is much easier to understand data in the visual form rather than in the form of text, numbers, and large tables with lots of row and columns. One can choose the data visualization technique wisely by understanding data and its composition.

Keywords Data visualization · Agile methodology · Line chart · Pie chart · Bar chart · Bubble chart · Symbol maps · Portfolio wall · Kanban board · Epic and story

1 Introduction

The world is throng with growing data on daily basis and there is a need to handle and display data in an understandable form. Visualization is a technique that is

P. Gandhi (✉)
Department of Computer Application, Manav Rachna International Institute of Research and Studies, Faridabad, India

J. Pruthi (✉)
Department of Computer Science and Technology, Manav Rachna University, Faridabad, India
e-mail: jyoti@mru.edu.in

S. M. Anouncia et al. (eds.), *Data Visualization*,
https://doi.org/10.1007/978-981-15-2282-6_4

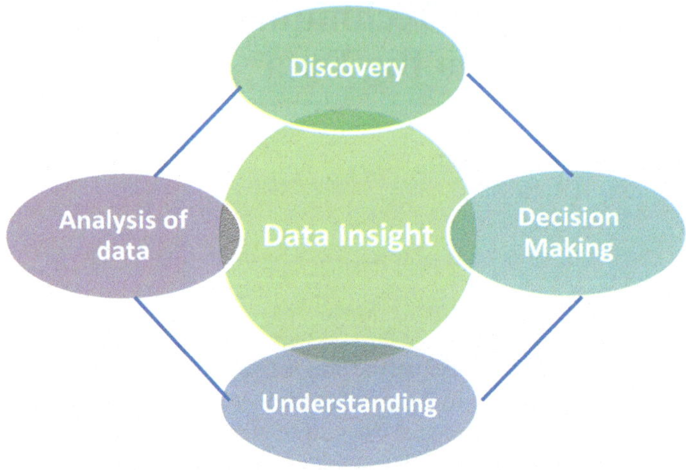

Fig. 1 Benefits of data visualization

used to analyze the data in variety of ways to make effective decision making. Data visualization is an effective process to present data and information graphically or pictorially. It is emerging like a powerful and widely applicable and acceptable tool for interpreting and analyzing large and complex data. It is becoming easy and quick way of conveying data along with concepts in a universal format (Fig. 1).

Data visualization is concerned with development, design, and application of graphical representation of data and makes it easy to understand the sense of data. It is also known as scientific visualization or information visualization.

Using pictures, graphs, charts, and maps, to understand the data and information have been used for centuries. Due to the advancement in computers, now it is possible to handle and process huge amount of data at a very high speed. Today data visualization is becoming a blend of art and science that is going to bring a visible change over the few coming years.

Visualizing data may be a challenge but it is much easier to understand data in visual form rather than in form of text, numbers, and large tables with lots of row and columns. One can choose the data visualization technique wisely by understanding data and its composition. All visualizations techniques are trying to solve the same problem but in a different way.

Broadly, there are two categories of data visualization with different purpose: explanation and exploration. Exploration data visualization is useful when data is available in quantity but knowledge about data is very little and goals are vague. Explanatory data visualization when again data is available in quantity but we know what exactly the data is. Both the categories help in the presentation of data visually.

This chapter provides an overview of data visualization, why it is important, factors involved in data visualization, different visualization techniques for big data, comparison between various techniques, related tools and software, Visualization for Agile software development, and selecting appropriate visualization technique.

2 Importance of Visualization

With extensive use of Internet and Web, a large amount of data is generated every day. There is a need to understand large and complex data. Each organization which keeps record basically deals with data and has to take decision. When the data is available in large volume, it has to be processed by using various data processing methods and need to present it with different types of techniques and methods.

It is a key to success of any enterprise as it helps enterprises to control the data in an effective manner and make the best utilization of that data to convert it into knowledge. It is a process of converting data and numbers into visual form. Data visualization techniques use different effects of computer graphics. Data visualization enhances learning, understanding, and reasoning and helps the stake holders to make effective and fast decision making. It also provides the better understanding for pattern recognition, analysis of trends, and to extract the appropriate information from the visuals.

Data visualization actually helps in communicating complex data with accuracy, clarity, and efficiency. It actually absorbs the data in a new and more constructive ways that help the organizations to take appropriate and useful decisions. It visualizes the relations and patterns among operational activities. In another words, we can say that data visualization is a new business language.

With the help of visual trends, one can easily understand what can be the best next step within very less time frame. It makes the data less confusing and more sharable and accessible. It would be easy to memorize and remember the data if it is in graphical format. Moreover, a very little modification can give the new look to the information and helps to formulate different strategies for different situations.

3 Factors Affecting Data Visualization

The successful visualizations have a few properties in common. They all included a specific and clear objective; they have only relevant information, focused data, and present the data in a way that projects the patterns and relations in the data.

One needs to be careful before adopting a specific visualization technique. To achieve the objective, there must be some points that need to be keeping in mind every time before going to visualize the data. Firstly, an objective of data visualization should be defined. Second, focused data should be selected. Third, a suitable technique must be identified. And after that, other options of colors, fonts, and other visuals can be chosen.

Fig. 2 Properties of data

If the purpose is clear and specific, it would be better to articulate the data in effective way. Before going for visualization, one should be clear about the stake holders of the end product. Always create the visuals that are valuable to the user (Fig. 2).

A good practice is to start with a question "What value the data visualization is going to contribute in decision making?" Data visualization with good quality comes in different sizes and shapes but all have some specific features that ensure to give something with proper insights of the data. Generally, a good visualization output must be meaningful, user uses it on regular basis and will be able to take effective decisions by comprehensive scenario; Desirable, it must be pleasant to use; Usable, user can use it to meet their objective very quickly and easily.

The visualization must be visually excellent appealing and the quality of the output should be good. The visualization should be scalable. As the data size would increase, the visualization application has to perform the same way. Therefore, the system should ensure the scalability for the future modifications.

4 Traditional Data Visualization Techniques

There are various tools and techniques which are used to convert the data in its visual form which cannot be directly converted by human being. Microsoft Word, Microsoft Excel, Microsoft Spreadsheet, and PowerPoint are some popular multipurpose tools with database connectivity as well to serve the purpose of data visualization and yields great results. These softwares are effectively used by the organizations which do not require highly specialized visualization of data.

Some of the traditional data visualization techniques to represent data are pie chart, line chart, bar chart, area chart, graphs, map, heat map, etc.

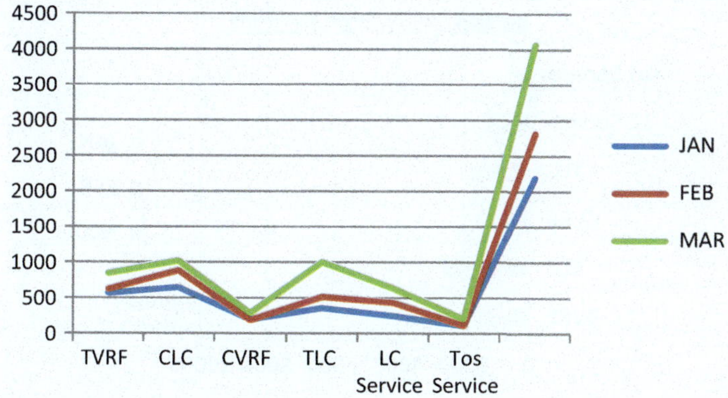

Fig. 3 Line chart for AC service in three months

4.1 Line Charts

It is one of the basic techniques to make the data more appealing and visualized. It shows the relationship between two patterns. It is also very effective to compare several values at the same time interval. It is the most effective approach when change in a variable or variables needs to be displayed (Fig. 3).

4.2 Pie Charts

It is also named as circle graph. The data is represented in the form of pie slice. The big slice shows the big amount of data. It is basically used to show the components percentage of the whole. Two popular variations of pie charts consist donut chart and exploding pie chart (Fig. 4).

Fig. 4 Pie chart for AC service

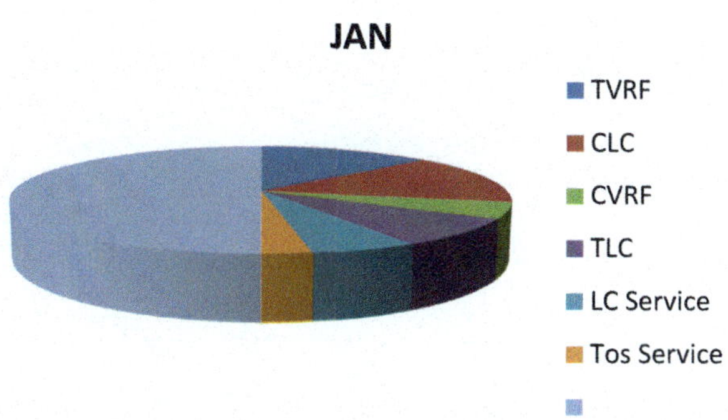

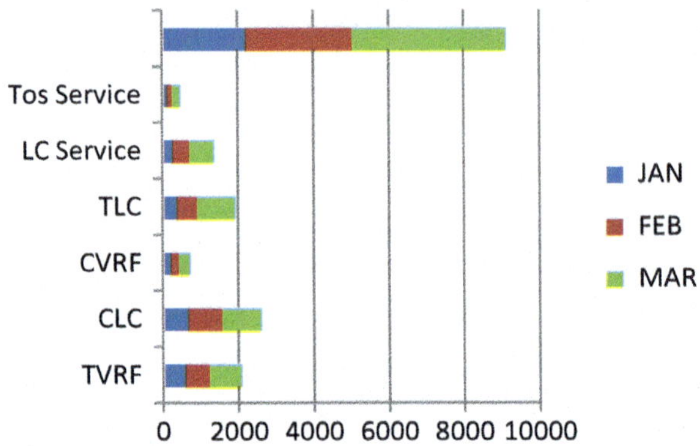

Fig. 5 Bar chart for AC service

4.3 *Bar Charts*

It is also referred to as column chart which makes the use of both horizontal and vertical bars. It is used to compare items of different groups. It is not very effective when the amount of data is very huge. It is mainly used by industries to compare their sales, cost, etc. (Fig. 5).

4.4 *Area Chart*

Area chart is the best choice when there is a need to show the trend over time. Line chart and area chart are similar in nature as in both the charts, the data points are plotted and connected through a line except that in area chart, the whole area between axis and line is filled in with color or shading (Fig. 6).

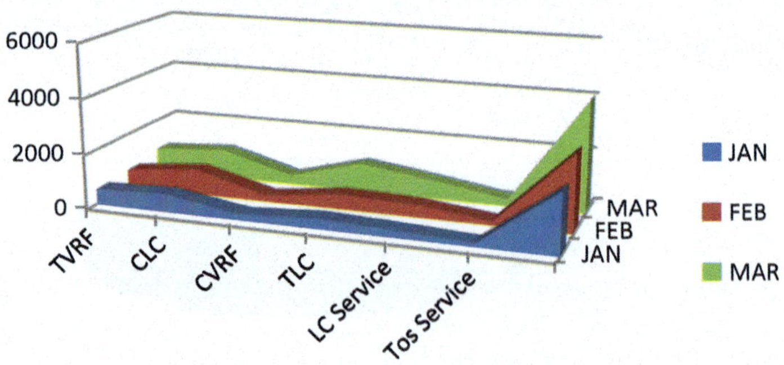

Fig. 6 Area chart for AC service

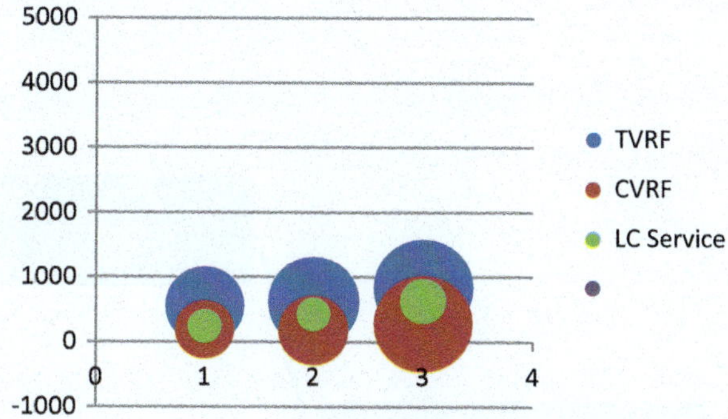

Fig. 7 Bubble chart for AC service

4.5 Bubble Chart

It is one of the variations of scattered plots in which the makers are replaced by the bubbles. It needs at least three measures, two for the plot axes and third for the size of bubbles to show the relationship. It is a good choice for the large set of data (Fig. 7).

4.6 Scattered Plot

It is a two-dimensional chart which is used to display the variation between two data items. A scatter plot is also called a scatter chart, scatter diagram, and scatter graph. It helps mainly to know how closely the data is related to each other by showing how the data points are scattered or spread over a graph area (Fig. 8).

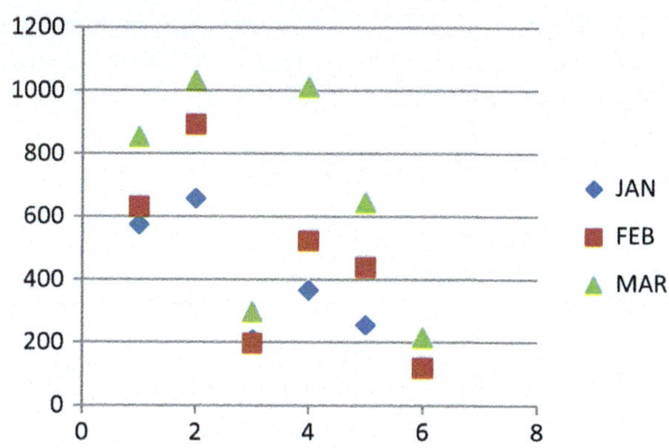

Fig. 8 Scattered plot for AC service

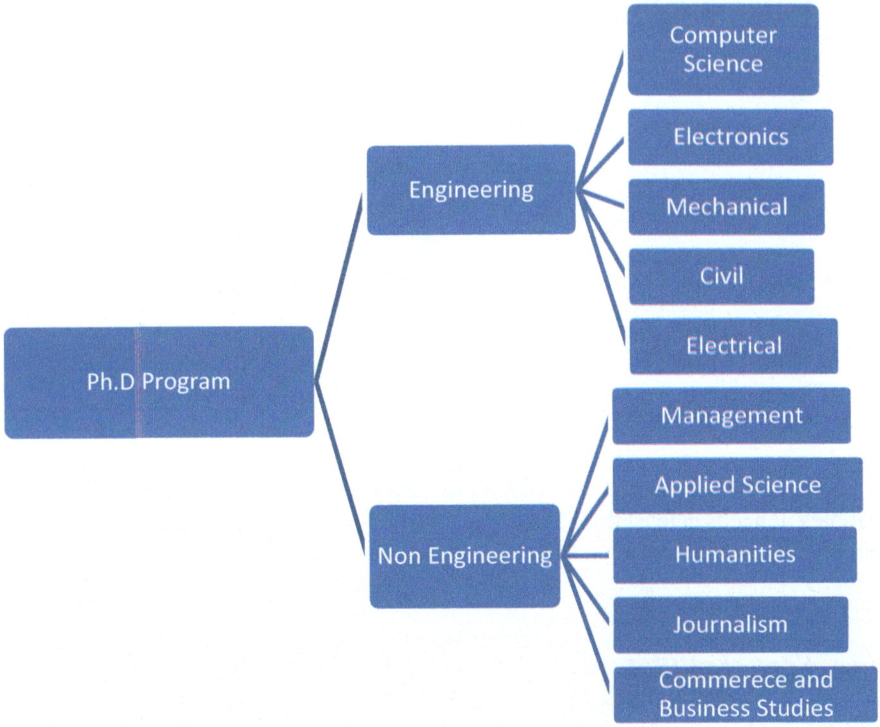

Fig. 9 Tree map for education system

4.7 Tree Maps

This approach is used to show the data in hierarchical form. The data is represented in the layered rectangular form to show the structure of hierarchies at different depth. The objects can be divided into various divisions and subdivision as per the requirement (Fig. 9).

4.8 Heap Maps

It is used to represent and compare the data using different colors. As an example, it can be used to show the best cases in green color, average cases in yellow color, and worst cases in red color which helps the end user to compare the performance of things in one go (Fig. 10).

These data visualization techniques suit both traditional as well as big data to some extent. These tools also save time to great extent as very less human intervention was

Fig. 10 Heap maps for AC service

	JAN	FEB	MAR
TVRF	578	635	858
CLC	660.42	897	1036.18
CVRF	212	201	302
TLC	370	527	1016
LC Service	260	444	650
Tos Service	123	124	221
	2204	2828	4083

there for visualization process and helps the researchers as well to conduct their research with the help of effective visual form of the data.

But when talking about advanced level visualization of huge amount of data, we require some specialized visualization tools, which give an entirely new insight to data and also have the capability to apply all permutation and combination to the variety of data and eventually results in the relevant and significant visual representation as compare to the traditional one.

5 Visualizing Big Data—Tools and Techniques

Visualization is not just a convenient but it is the one of the prominent features for big data. It is a big challenge to handle variety of big data as each data has different speed, size, and diversity that should be taken into account in order to visualize. There are three vs that affect the operation of data and must be taken into account while doing data processing and visualization. These three vs are volume, velocity, and variety of data.

Volume refers to the size of the data that is accessible to any organization and can be in terabytes, petabytes, etc. Variety refers to the representation of data in audio, video, text, and images forms and it also refers to structured and unstructured data. Velocity means the frequency of changing the data. It also refers to the factoring and aggregating the data.

The modern data visualization techniques to represent and handle big data are word clouds, symbol maps, connectivity chart, etc. These techniques are specially designed to handle semistructured and unstructured data.

5.1 Word Clouds

Word cloud is a technique for visual representation of text data. It is useful for analyzing sentiment analysis of the post done by people in social media. It highlights the most frequently used keywords on a Web page. The importance of each word is indicated by using different font size or color. This technique helps in finding the

Fig. 11 Word cloud for frequent terms

most prominent words in a quick manner. Word clouds are the most widely used technique due to its readability, understandability, and simplicity. It can be easily shared and are very impactful. Word clouds are used by almost all fields of life. It can be used by researchers, marketers, educators, politicians, journalists, and social media sites.

Tool
Many freeware softwares are available which helps to process text given by user and results in the significant word clouds. Tagul is one of the tools that can be used to create word cloud based on the text already available and new text imported by the client or user. The resultant data of Tagul tool is highly customizable and animated with the features to specify different color, shape, and size (Fig. 11).

5.2 Symbol Maps

It is same as word cloud except that we represent symbols in place of words. The symbols will be of different sizes which make them easy to compare. To create a symbol map, there is requirement of quantitative value or location names. The large variation in the data is advisable to see the difference in symbols, otherwise all the symbols will appear of same size and will be difficult for business user to distinguish them and find the optimized results of visualization.

Tool
Many freeware softwares are also available which helps to process text given by user and results in the significant symbol maps.

Tableau v4.0 is the most effective tool for creating symbol map. Before this, people have daunting experience to create symbol maps. Tableau v4.0 makes the symbol

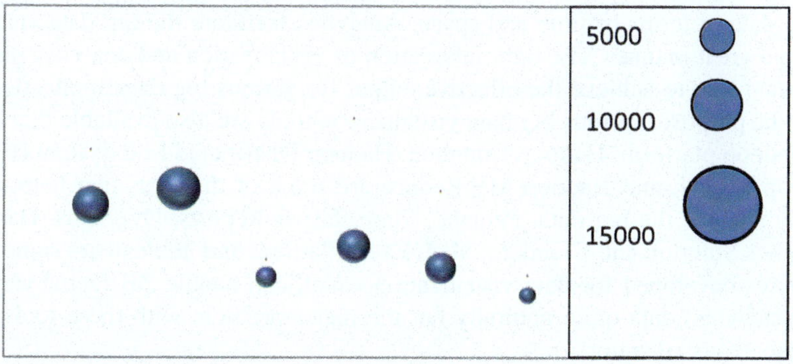

Fig. 12 Symbol maps

maps more effective for visualization purpose and also helps the business users to categories data for decision making (Fig. 12).

5.3 *Connectivity Charts*

This visualization technique is used to show the connection between action and their triggers. It also shows the strength of connection between them (Fig. 13).

Effective visualization of huge data is not possible without analytics. In order to achieve the optimization, the data should be preprocessed to lessen the complexity

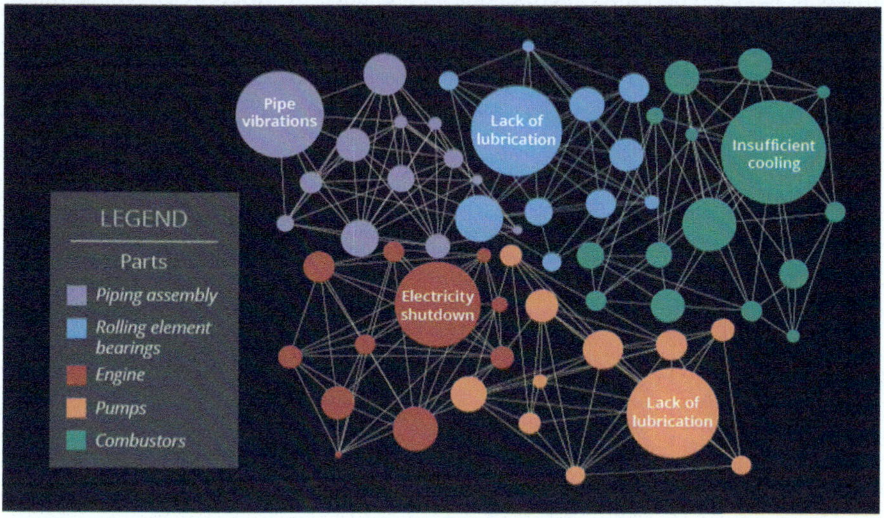

Fig. 13 Connectivity charts

of big data in terms of time and space. Analytics facilitate the big data visualization in a great manner. The tight integration of visualization and analytics plays an important role to achieve the effective output for various big data applications. To serve the purpose, various big data visualization tools are also available that run on the Hadoop platform. Hadoop Common, Hadoop Distributed File System (HDFS), Hadoop YARN, and Hadoop MapReduce are some of the tools that help to efficiently analyze the big data. Pentaho, Flare, Jasper Reports, Dygraphs, Datameer Analytics Solution and Cloudera, ManyEyes, Platfora, and Tableau are some of the softwares developed for data visualization which can handle ZB (zettabytes) and PB (petabytes) data quite naturally but the major problem with these tools is the inadequate visualization.

6 Visualization in Agile Software Development

In Agile, analysts help to define a specification and it is all about division of work among teams. More interaction would be there between team members and teams. Data visualization allows organization to see the progress of teams to achieve the objective. The traditional way to do the same is "Waterfall Life Cycle" (Fig. 14).

Waterfall life cycle process is bit lengthy and time consuming too. There is a lot of time gap between the requirement gathering and completion of project. The role of the consumer is only at the beginning and at the end. There would not be any communication in between the first and the last stage. There may be a chance that requirement could be change because of long time gap.

Moreover, there would be a very less interaction among the team members of different stages. Therefore, there could be a chance of different understanding of

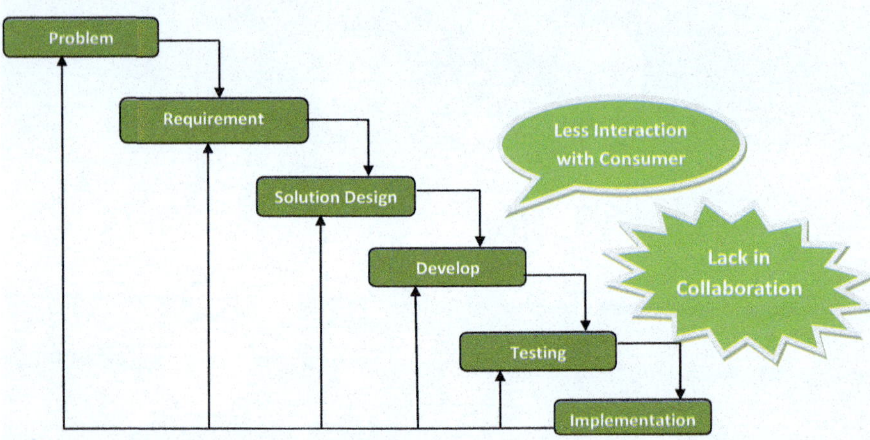

Fig. 14 Working of waterfall model

different teams for a same problem statement. Developer would not be in touch with consumer. So there is always a gap between what is expected and what is delivered.

To overcome all the above-mentioned issues, an iterative model can be introduced where in more team collaboration and more consumer interaction would take place. Agile software development is one of the solutions for the same. It allows the quick delivery, quick feedback, quick review, and changes can be done which result in desirable output and ultimately meeting the expectations of the consumer. It also has the ability to fetch data from different sources and display data by using visualization technique (Fig. 15).

In Agile, there is daily stand-up meeting to fix the target for the day and different iteration would be there as per the communication and feedback of the consumer on the deliverables. Sprint planning and release would be done on regularly basis as per the requirements and lastly, strategy would be formed to make output a better version (Fig. 16).

In Agile, to show the progress and the status of work done, on daily basis is presented through some visualization techniques. These techniques are very different, powerful, and useful to convey the progress. Some of the techniques are

(a) The portfolio wall
(b) The kanban Board

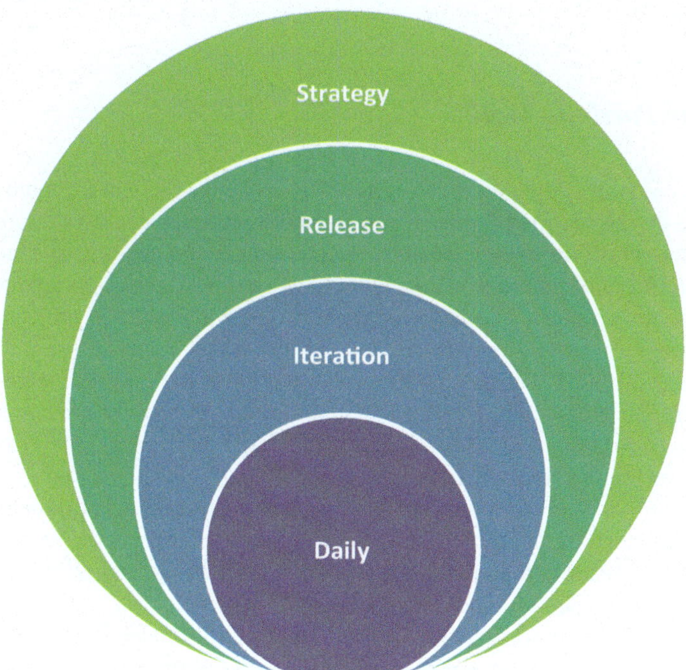

Fig. 15 Agile methodology

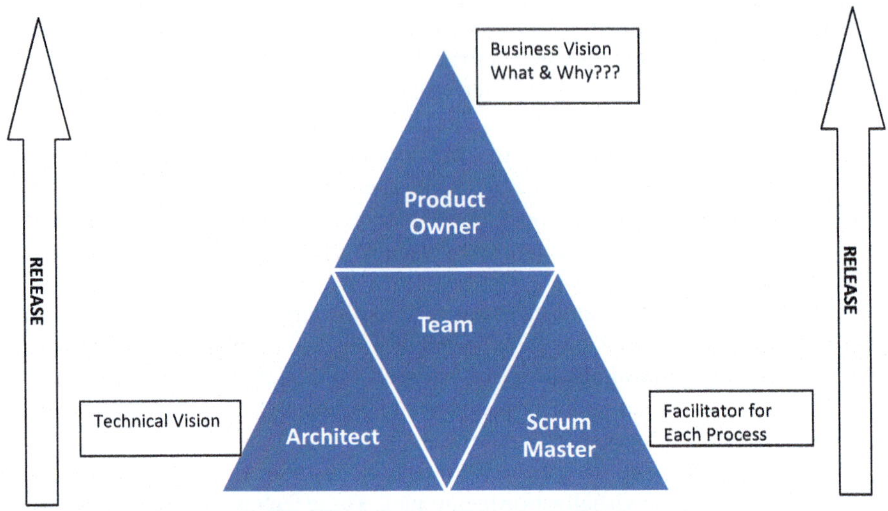

Fig. 16 Collaborative way in Agile

(c) Burndown chart
(d) Epic and story Mapping.

6.1 The Portfolio Wall

The portfolio wall is the very powerful tool to visualize the work done by the team to achieve the pre-defined targets. It tracks and treats each cycle as "Iteration." All the completed tasks remain in the current iteration and the pending one moves to the next iteration. This technique allows team to work toward the combined release and to visualize the progress. In other words, it will create a visual that projects the status of milestone achieved till now.

The portfolio wall uses the color codes to represent the different teams and their integration. Each team has its own color scheme and shows their milestone that they have to meet for the successful release. It increases the transparency across teams and removes the deadlocks (Fig. 17).

A portfolio wall takes a matrix-like structure with time on horizontal axis and goals on vertical axis. Card on each column actually represents the work to be done by a particular team in a definite time. Lines, colors, cards, and other attributes are used to represent team and their dependencies.

Nominated team members from each team generally meet every day in front of the wall in order to monitor and discuss the progress of respective teams.

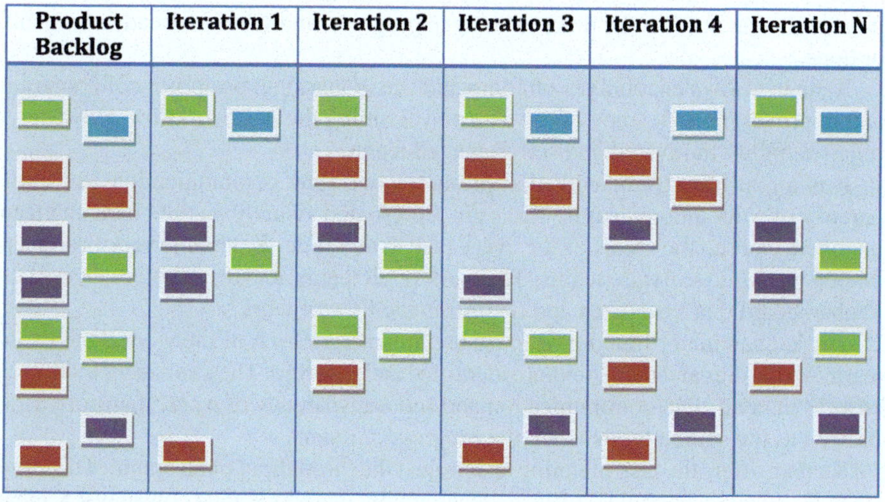

Fig. 17 Product backlog-task and associated user stories

When a team creates a plan in collaboration and it will be displayed on the wall, their engagement increases, and improves the output. There are a number of benefits of using the portfolio wall:

1. By making an active participation in maintaining and creating the plan for work to achieve common objective, the team sees some control over the work.
2. A portfolio wall allows the team to visualize the big and clear picture of their work progress.
3. The target seems more achievable when items move from one section to another.
4. The wall is visible to everyone that increases the level of commitment of team members.
5. The portfolio wall actually encourages the team members to work harder to achieve the goals and meet the objective.

Some of the major disadvantages of the portfolio wall are:

1. The historical data cannot be projected on the wall.
2. There is a lack of visibility in projecting unallocated work but actually needs to be done.

6.2 The Kanban Board

Kanban is a method or a technique that is used in Agile to manage the creation of products with an emphasis on continuous delivery but at the same time, not overburdening the team. It is a process to help teams to work effectively in collaborative manner. In this process, each member is able to see the progress and always be informed what needs to be done, what is in progress, and what is already completed or done.

Kanban always encourages ongoing and active learning, promotes collaboration, and improves the efficiency of the teams by defining the best workflow. However, it requires highly motivated and self-managed teams.

It is a real-time framework that promotes real-time communication and transparency of work among teams. Tasks are represented visually on the board and team members can see the status of the every task at any time. Kanban is very prominent among the Agile software teams. The work of all teams actually revolves around the kanban board that visualizes and optimizes the flow of work.

The kanban major function is to ensure the visualization of the work done by the team. All hurdles and deadlocks are identified and resolved. The kanban methodology ensures the real-time communication and full transparency of work, therefore it can be seen as the source to visualize the progress of teams.

Kanban offers the task planning and project the throughput of all teams of different sizes. It is the most popular software development methodologies nowadays. Kanban helps to visualize who is responsible for what. It increases the focus of the team to complete the work and achieve the objective (Fig. 18).

The Kanban board is basically divided into three sections

1. To-Do
2. In-Progress
3. Done.

The task that needs to be performed at starting comes under the To-Do column. The task on which team is currently working on comes under the column In-Progress. The task completed by the team to achieve the target comes under the Done category. Therefore, with the help of visualization through Kanban technique, all teams come to know about their regular performance and the task done by the team within a limited time frame. The entire team gets focus on completing the tasks that are in

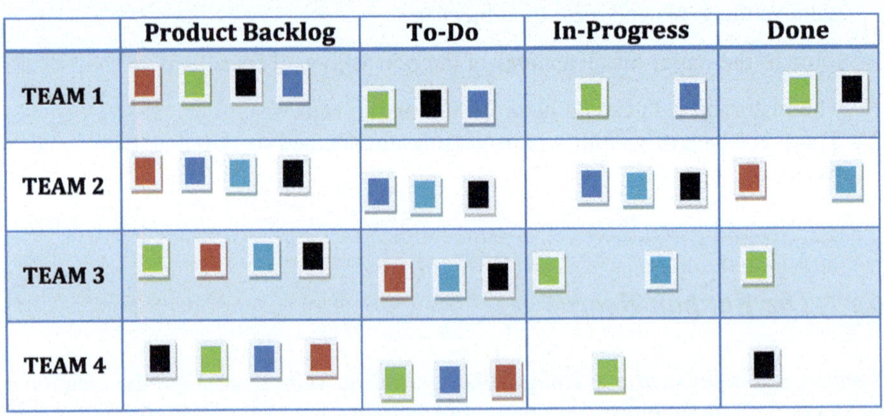

Fig. 18 Team-wise progress record and checkpoints working of waterfall model

progress. The main idea behind the Kanban wall is to promote "**Stop Starting, Start Finishing.**"

The major benefits of the Kanban are:

1. It helps to shorter cycle time and faster the delivery of the features.
2. The team can visualize the performance on daily basis and they are ready to adapt the new strategy to achieve the target and get the delivery done within the time limits.
3. Kanban is ideal where priorities are changing very frequently.
4. The feedback is projected on the wall that improves the chances of more motivation in a team and more empowered performing team members.
5. Kanban provides the better transparency for each task that needs to be performed.

Some of the drawbacks of the kanban are:

1. The outdated boards can mislead the team at development stage. The wrong or duplicate issues will float to the development process.
2. There may be a chance that teams make the overcomplicated kanban boards that result in wastage of time in understanding the board instead of finishing the work.
3. Sometimes, there is no time associated with each phase that leads the team in wrong direction.

Most of the kanban disadvantages are due to mishandling of kanban board. Therefore, proper understanding of how to use kanban board is very necessary.

6.3 The Burndown Chart

The Burndown chart is a graph that represents the progress of team over time while doing the project. It tracks daily progress of the team and is considered as an essential tool to track the progress. When the task is completed, the graph burns down to Zero (Fig. 19).

The burndown charts represent the relation between the amount of work (x-axis) and the time (y-axis). Time is shown with days when the work started but in case of Agile, it may be represented in terms of sprit also. The vertical axis represents the wok left to complete the task is the sum of estimation.

The burn down chart has two lines:

- Estimated task (Defined)
- Actual task (Deviation).

Estimated task (Defined) is a straight line, starting from the start point up to the end finishing point. It is the actual path that needs to be followed by the team to achieve the set targets. This line represents the tasks need to be completed in one day. There may be a chance that the team takes longer or short time span to complete the defined tasks. Actual Task (Deviation) is the actual work at given point of time. This is the line that shows the actual progress of the team or in other words, it shows

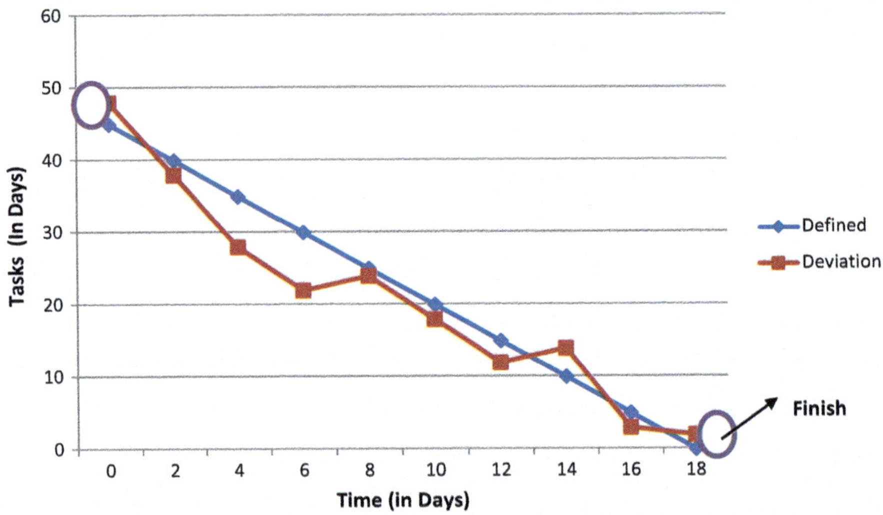

Fig. 19 Burndown chart—progress tracker

the deviation from the actual estimated. The team can understand the deviation from the defined path on daily basis.

The reading of burndown chart is very important as it shows the actual performance. When the actual task or deviation line is below the defined line, it means that the team is ahead of schedule and completed the task before the estimated time. In contrast, when the actual task line is above the defined line, it shows that team is behind the time schedule and is not able to complete the task in defined period of time.

The balance between both the lines is very important. It also helps in identifying the potential of the team and helps the team to so estimation in appropriate manner so that the objective can be meet as per time frame.

Big visible charts play very important role and are considered as a powerful tool to visualize the performance of the teams. It makes the process and progress transparent to everyone. It also helps the team to identify their strength and weakness through retrospective and, therefore, the team can work on the items that need to be improved.

Some of the major benefits and drawbacks of burndown charts:

1. Burndown chart forces team to evaluate their performance on daily basis and make the strategy as per retrospect.
2. It helps to maintain the accuracy of sprint and ensure the timely delivery of work.
3. Sometimes, it is difficult to maintain the burndown chart and one wrong projection on the chart may lead to break down the efficiency of the team.

6.4 Epic and Story Mapping

Story mapping is a technique introduced by Jeff Patton. It involves building a grid of user stories that is actually representing the experience of the users. User story arranges user stories according to how people think and do work on daily basis. It represents the actual thinking and real experiences of the users and tries to make strategy to proceed for work and ultimately draws conclusion what is to be done and how it is to be done.

The main aim to create a story map is to ensure that all teams and their members are on the same page from the start of the development and commit to achieve the same objective. The story map is considered as a collaborative practice that guides the team to create the product backlog and visualize the target.

The first step to create the story map is to decide the flow of activities by the user. This should be considered as the core flow of the user activities. The team should create the set of activities and need to arrange the same in chronological order.

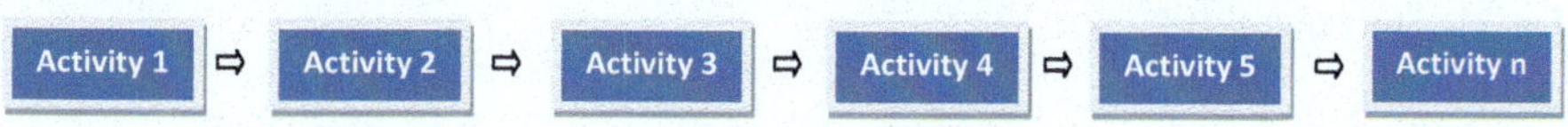

After creating the pool of activities and arranging them in order, the second step is to identify the task with each activity and group them with respective one. The task has to be identified keeping in mind the time needed to complete the activity. The task should be associated with activity.

There may be possibility the number of tasks may vary from activity to activity. It is not necessary that each activity should have equal number of tasks under it (Fig. 20).

The next step is to map the user stories with corresponding activities and task. The mapping of user story needs to be done very carefully as per the requirement and the previous experiences of the user.

The format to write a story is:

As a user, I want (define goal), so that I can achieve (mention objective or reason) (Fig. 21).

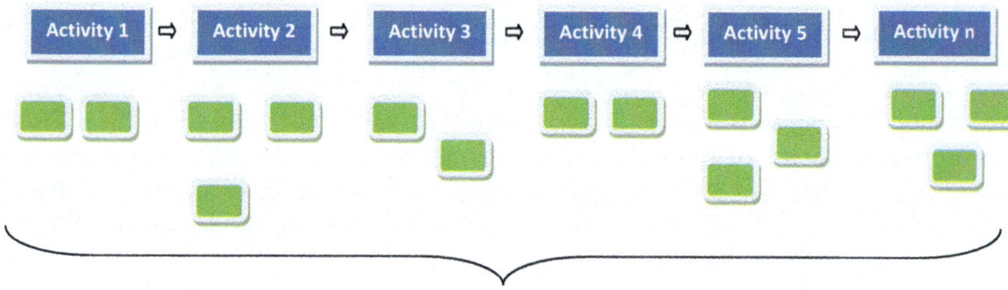

Fig. 20 Tasks associated with each activity

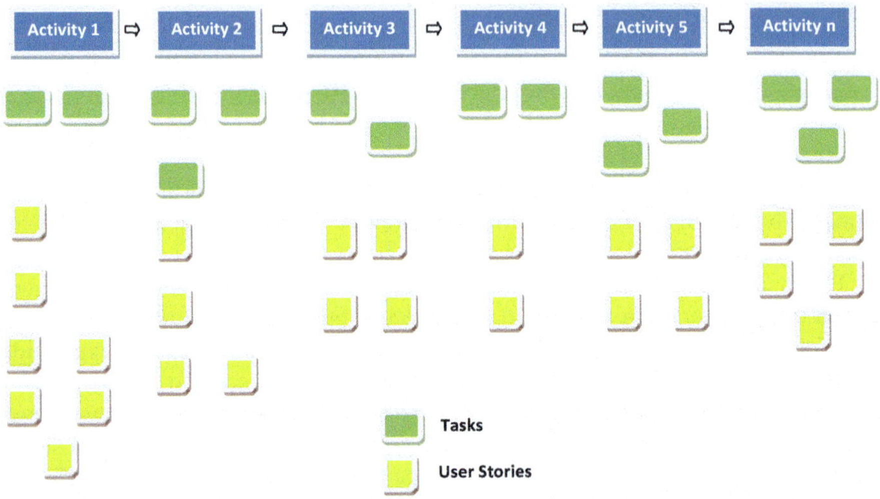

Fig. 21 Story framework

The story mapping is considered as the backbone that map and convert the user stories into backlog and create the visual to achieve the objective. In this way, a product can be developed that value the user and meet the expectations of the user in mass.

Story maps can be seen as the great information radiators. But it requires great space to capture the entire stories.

Looking at all the major visualization techniques under Agile software development, the team has to understand what to choose and when to choose. As these visuals can do a paradigm shift and are able to introduce more efficient and productive processes to achieve the objective within a prescribed time frame.

All methodologies have some pros and cons. So select it very carefully and as per the requirement of the organization. The combination of two or three visualization techniques can also be used and it may prove to be more effective technique.

7 Challenges of Big Data Visualization

Two major challenges of big data visualization are scalability and dynamics. Speed is a prominent factor of big data visualization. Designing the visualization tool is a very tedious task for efficient visualization. Parallelization is another challenge of visualization due to the size of huge amount of data. While doing the big data visualization, the following problems can be faced:

- Loss of information
- Noise recognition

- Image sensitivity
- High performance requirement
- High rate of image change.

8 Choosing Appropriate Visualization Method

Obtaining optimized visualization of data totally depends on how effectively you find the best underlying visualization method that suits the requirement as well as displays the data appropriately. Because at different times different visualization techniques must be used to accomplish variety of tasks. This is a big challenge these days as most professionals still do not know which is the best technique to use to achieve a goal or accomplish a task and eventually end up getting wrong results for correct data. Therefore, it is quite interesting and challenging as well to find out the best one for your code and end up with the optimized results.

Both the above-explained visualization techniques for big data and traditional data can be used for big data. The only thing matters that it must result in easily understandable output that helps the business users to take wise decision. SAS visual analytics is one of the important approaches that enable us to explore the data using various visualization techniques. Exploration data visualization is useful when data is available in quantity but knowledge about data is very little and goals are vague.

Box plots and correlation matrices are such techniques that help us quickly understand the data and its composition irrespective of its size. SAS visual analytics works in two phases: At the front end, it helps large number of users to view and interact with the report to take further decision, while at the back end, it concern is also available which ensures the security of the underlying data and also controls other aspects as well directly or indirectly related to the security of data.

This results in the fast-track processing of data, makes it available to the hands of decision makers, and makes them more productive and collaborative with the optimized results.

9 Conclusion

To understand the information in visual form is much more easy as compared to the information in the traditional manner like in the form of table, text, etc. Visualization is an area which makes it easy to interpret the data and go for corrective decision and its composition. Visualization techniques are available to handle both traditional and big data. Various visualization choices are available but some of the techniques may end up with the wrong visualization presentation. Thus, it is important to choose the appropriate visualization method to better understand the data for further business analysis and many more. In this chapter, we tried to show all the alternatives available that conveys the data more clearly as well as truly understand the data.

10 Summary

Data visualization is one of the interactive ways that leads to the new innovation and discovery. It is a dynamic tool that opens new ways of research which facilitate the scientific process. The main objective of this chapter was to explain:

- Factors affecting the data visualization
- Visualization techniques for traditional data
- Visualization techniques for big data
- Tools and software available for visualization
- Visualization for Agile software development
- Choosing appropriate visualization technique.

Data Visualization: Visualization of Social Media Marketing Analysis Data to Generate Effective Business Revenue Model

Aditya Chellam, Ayush Chaturvedi and L. Ramanathan

Abstract Right from its inception, social media has played a pivotal role in shaping the marketing strategies of today's business. Businesses use marketing to successfully grow their market presence and improve brand awareness. The most effective marketing approach is one where social media and traditional marketing mixes are used in tandem. Social media marketing is a lucrative option for business owners as the cost of marketing is low and user feedback on social media Web sites and forums can be utilized effectively to constantly update the marketing strategy for maximizing gains. This chapter focuses on analyzing the Facebook marketing strategy of a certain company and providing a comparative study of visualization methodologies that present the client sentiment in the most lucid manner, thereby allowing the business owner to devise an effective business model with maximum returns and minimum expenditure.

Keywords Data visualization · Social media · Market analysis · Facebook advertisement · Click-response · Business expansion

1 Introduction

Social media has become a power to be reckoned with in the modern marketing and promotional domains. It is thus vital that company marketing managers acknowledge its power of shaping public opinion in making purchase choices. Consumer-to-consumer interaction lays the foundation on which public perception of a brand image is based on and thus justifies the hybrid component of social media which combines consumer-to-corporate interactions with a direct consumer-to-consumer interaction. The ease and availability of social media have made it the primary source of information and feedback for consumers and companies alike. It gives more control on content review in the hands of the consumers. This content is unsupervised and cannot be controlled directly by the organizations. Thus, marketing managers should make use of various strategies to shape these conversations by appealing to

A. Chellam · A. Chaturvedi · L. Ramanathan (✉)
VIT University, Vellore, India
e-mail: lramanathan@vit.ac.in

S. M. Anouncia et al. (eds.), *Data Visualization*,
https://doi.org/10.1007/978-981-15-2282-6_5

the consumers' psychological needs so that they are predisposed to give a positive response to other potential buyers on social media forums [2]. Data visualization is an enabling means that allows marketing managers to effectively visualize marketing perceptions [1].

Various social media platforms allow users to share their views on a product or a service in different ways, for example, YouTube and Vimeo allow for sharing video-based reviews, blogs and forums allow for textual reviewing (Blogger, WordPress, Reddit, and Twitter), and Web sites like Facebook and Instagram allow for a combination of all of these forms. This allows a user (customer) to express opinions in more memorable and impactful formats; a segment which has been discussed in quite some detail at the end of the paper. In the earlier marketing strategy, the frequency, content, medium, context, and timing of the promotion were entirely controlled by the company management. Social media has altered this vertical hierarchy and replaced a part of it with a more flat hierarchy, by allowing customers to interact with one another. Thus, now, the control of what is being talked about the product is in the hands of the customer. There is no restraint on the content, frequency, or any of the aforementioned points as was the issue in the erstwhile paradigm. Social media has also altered the schema of issue redressal. Now, customer-to-customer interactions can affect the purchasing decisions of other customers. Thus, companies have to be more cautious and vigilant about how they handle their mistakes.

2 Background

Citing the work of Ricadela, in 2007, who paraphrases the words of the co-founder and chairman of LinkedIn, Reid Hoffman, "the ability to leverage relationships embodied in social networks will become one of the most transformative uses of the Internet." Social media has extended the domain of marketing beyond the geodemographic spheres of traditional marketing means, as various means of content consumption can be availed by users in the form of blogs, articles, and videos. According to the study, a Web-ranking analytics company, the top 10 Web sites in 2010 accounted for about 75% of total page views in the USA, which is a steep increase from the 31% in 2001 and 40% in 2006. Many businesses such as Whole Foods, Zomato, Zappos, and Swiggy actively connect with consumers on a variety of social networking sites.

Demographics are the key to any marketing strategy, and on social media, it is no different. When looking at Facebook, there are nearly 1.15 billion people scrolling through their feeds every day, so it is important to know who need to be reached and how. Some of the factors to be kept in mind for a business are age, gender, location, and income. For instance, the fact that women tend to adopt Facebook more frequently than men and the core age group is 18–29 if visualized effectively can allow a business to plan their marketing mix to elicit maximum product acceptance in their respective demographics. Facebook has a wide ranging audience that extends across a majority of the globe with the income levels of the users spread across quite a wide range. Despite this widespread audience, it is observed that in the USA, most of the urban areas have a significant presence on Facebook, approximately 80%, and suburban areas constitute around 77%.

Businesses can get a significant boost in developing an effective revenue model by visualizing the ecosystem and assessing the market in terms of media ownership and the factors that influence different categories of social influences [8]. Doing so would allow business owners to determine and monitor key performance indicators (such as click to follow-up ration) and to define the key outcomes associated with that specific ecosystem in terms of appropriate metrics. A unique visualization would provide means to customize user engagement such as the use of the Fanbuzz Visualizer in Grammy Awards as part of the "We're All Fans" campaign significantly improved ticket sales leading to the eventual success of the Grammy Awards in 2010.

The objective of this chapter is to elucidate the marketing strategy employed by the company in areas of diversified interests, by monitoring the client click-response impact parameter. This chapter also attempts to categorize market diversification via a numeric representation of an estimated number of people that show interest in buying the products.

3 Purpose

This study primarily focuses on investigating the advertising strategy employed by the company in areas of diverse interests and to record the type of response they receive in the form of clicks. The goal of the study is to compare the money spent on these campaigns, the clicks received on their advertisements, and the response of the customers in the form of actually ordering the products, which determine the efficacy of each advertisement on Facebook.

In this study, the market diversification has been categorized via numeric representation of an estimate of the people who show interest in buying these products. Accordingly, the following hypothesis has been constructed:

Hypothesis: *The Sales of any product increase when the click rate, frequency of advertisements and Money Spent for that product increases from the customers who express interest in that field.*

4 Dataset Description

Collected data from the company sales statistics have been used as the dataset for the purpose of this study. The file conversion_data.csv contains 1143 observations in 11 variables. Below are the descriptions of the variables.

(1) ad_id: *a unique ID for each ad*
(2) xyz_campaign_id: *an ID associated with each ad campaign of XYZ company*

(3) fb_campaign_id: *an ID associated with how Facebook tracks each campaign*
(4) age: *age of the person to whom the ad is shown*
(5) gender: *gender of the person to whom the ad is shown*
(6) interest: *a code specifying the category to which the person's interest belongs (interests are as mentioned in the person's Facebook public profile)*
(7) Impressions: *the number of times the ad was shown*
(8) Clicks: *number of clicks on that ad*
(9) Spent: *Amount paid by company xyz to Facebook, to show that ad*
(10) Total conversion: *Total number of people who enquired about the product after seeing the ad*
(11) Approved conversion: *Total number of people who bought the product after seeing the ad.*

5 Methodology

5.1 Importing the Dataset and Initial Analysis

The following are the steps to describe the dataset and for performing rudimentary analysis using statistical metrics,

5.1.1 Pseudocode

1. *Set working directory to dataset csv*
2. *Read dataset*
3. *Attach sales dataframe*
4. *Convert to string.*

5.1.2 The Dataset Description

See Figs. 1 and 2.

5.2 Constructing the Correlation Matrix and Corrgram

The following is the pseudocode for generating the correlation matrix and corrgram,

```
## 'data.frame':    1143 obs. of  11 variables:
##  $ ad_id                : int  708746 708749 708771 708815 708818 708820 708889 708895 708953 708958 ...
##  $ xyz_campaign_id      : int  916 916 916 916 916 916 916 916 916 916 ...
##  $ fb_campaign_id       : int  103916 103917 103920 103928 103928 103929 103940 103941 103951 103952 ...
##  $ age                  : Factor w/ 4 levels "30-34","35-39",..: 1 1 1 1 1 1 1 1 1 1 ...
##  $ gender               : Factor w/ 2 levels "F","M": 2 2 2 2 2 2 2 2 2 2 ...
##  $ interest             : int  15 16 20 28 28 29 15 16 27 28 ...
##  $ Impressions          : int  7350 17861 693 4259 4133 1915 15615 10951 2355 9502 ...
##  $ Clicks               : int  1 2 0 1 1 0 3 1 1 3 ...
##  $ Spent                : num  1.43 1.82 0 1.25 1.29 ...
##  $ Total_Conversion     : int  2 2 1 1 1 1 1 1 1 1 ...
##  $ Approved_Conversion: int  1 0 0 0 1 1 0 1 0 0 ...
```

Fig. 1 Dataset overview

```
##      ad_id          xyz_campaign_id fb_campaign_id      age       gender
##  Min.   : 708746   Min.   : 916    Min.   :103916   30-34:426   F:551
##  1st Qu.: 777633   1st Qu.: 936    1st Qu.:115716   35-39:248   M:592
##  Median :1121185   Median :1178    Median :144549   40-44:210
##  Mean   : 987261   Mean   :1067    Mean   :133784   45-49:259
##  3rd Qu.:1121805   3rd Qu.:1178    3rd Qu.:144658
##  Max.   :1314415   Max.   :1178    Max.   :179982
##     interest        Impressions          Clicks            Spent
##  Min.   :  2.00   Min.   :      87   Min.   :  0.00   Min.   :  0.00
##  1st Qu.: 16.00   1st Qu.:    6504   1st Qu.:  1.00   1st Qu.:  1.48
##  Median : 25.00   Median :   51509   Median :  8.00   Median : 12.37
##  Mean   : 32.77   Mean   :  186732   Mean   : 33.39   Mean   : 51.36
##  3rd Qu.: 31.00   3rd Qu.:  221769   3rd Qu.: 37.50   3rd Qu.: 60.02
##  Max.   :114.00   Max.   :3052003   Max.   :421.00   Max.   :639.95
##  Total_Conversion Approved_Conversion
##  Min.   : 0.000   Min.   : 0.000
##  1st Qu.: 1.000   1st Qu.: 0.000
##  Median : 1.000   Median : 1.000
##  Mean   : 2.856   Mean   : 0.944
##  3rd Qu.: 3.000   3rd Qu.: 1.000
##  Max.   :60.000   Max.   :21.000
```

Fig. 2 Initial analysis and metrics

5.2.1 Pseudocode

1. *import library(corrplot)*
2. *import library(corrgram)*
3. *columns = sales.df[,c("interest","Impressions","Clicks","Spent", "Total_Conversion","Approved_Conversion")]*
4. *n = cor(columns)*
5. *plot corrplot using method = "circle"*
6. *corrgram(columns, upper.panel = panel.pie).*

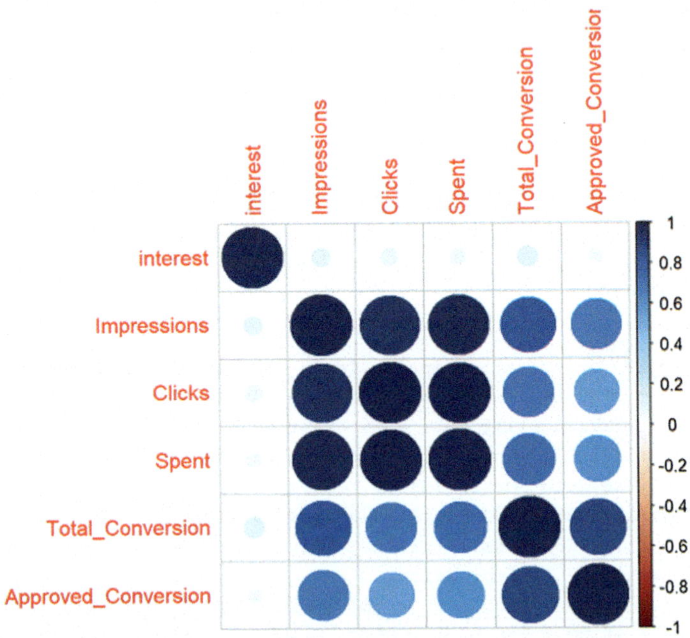

Fig. 3 Corrplot visualization

5.2.2 Corrplot Visualization

See Fig. 3.

5.2.3 Corrgram Visualization

See Fig. 4

5.2.4 Scatterplot Visualization

See Figs. 5 and 6.

5.3 Hypothesis Testing and T-Test

Hypothesis H1: *The sales of any product increase when the click rate, frequency of advertisements and money spent for that product increases from the customers who express interest in that field.*

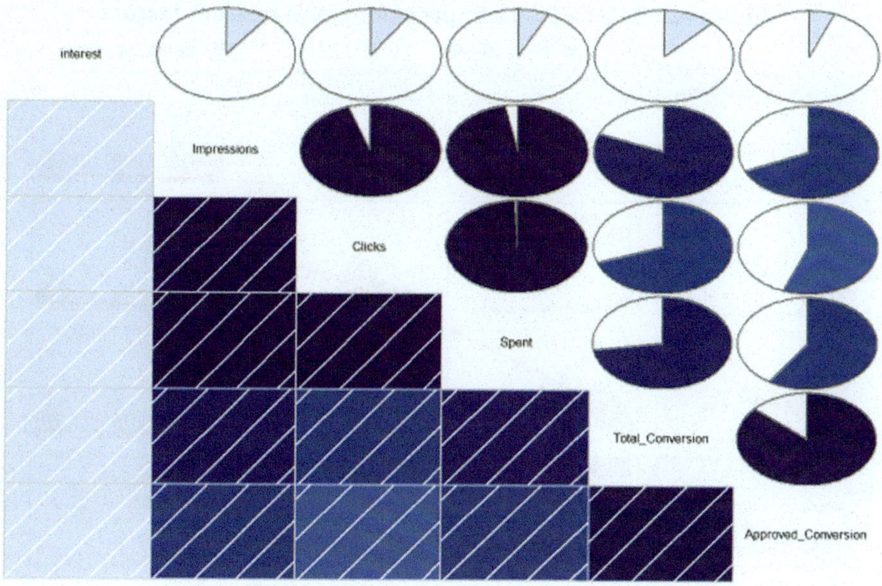

Fig. 4 Corrgram visualization

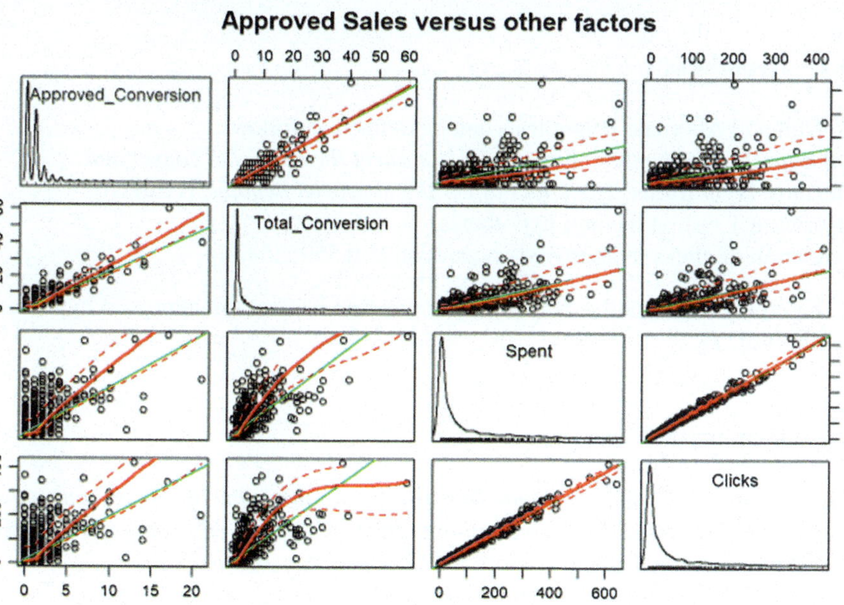

Fig. 5 Scatterplot of approved sales against other factors

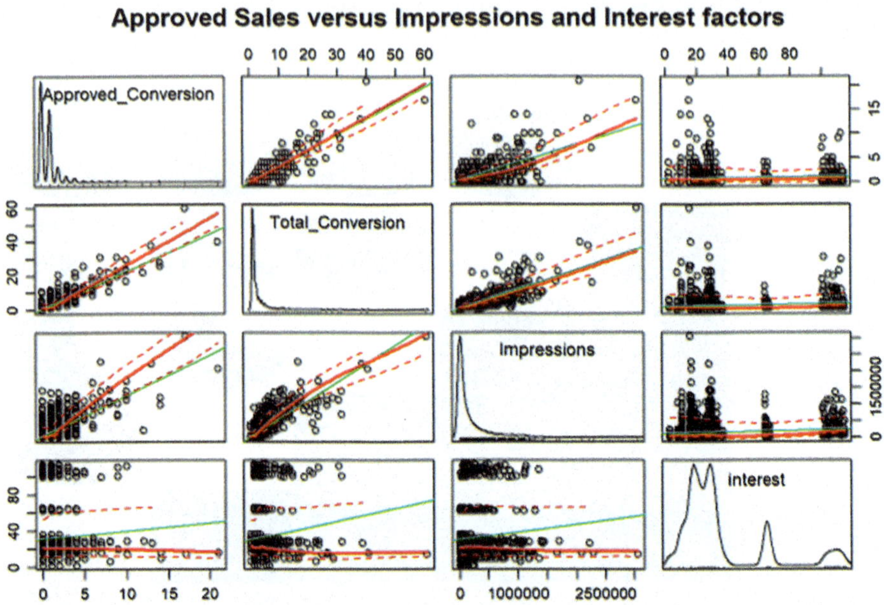

Fig. 6 Scatterplot of approved sales against impressions and interest factors

5.3.1 Hypothesis 1

The Null and Alternate Hypothesis can be defined as follows,
Null hypothesis: *Number of clicks did not affect the approved conversion.*
Alternate Hypothesis: *The clicks affect the approved hypothesis directly.*
Performing T-test on the null hypothesis,
Correlation of clicks to approved conversion = 0.5595258

Therefore, null hypothesis is rejected; the clicks affect the approved hypothesis directly (Fig. 7).

5.3.2 Hypothesis 2

Null hypothesis: *Amount of money invested did not affect the approved conversion.*
Alternate Hypothesis: *The spent amount affects the approved hypothesis directly.*
Correlation of amount spent to approved Conversion = 0.5931778
Performing T-test on the null hypothesis,

Therefore, null hypothesis is rejected; the amount spent is a good investment as it affects the approved hypothesis directly (Fig. 8).

```
##
##  Welch Two Sample t-test
##
## data:  Clicks and Approved_Conversion
## t = 19.272, df = 1144.1, p-value < 2.2e-16
## alternative hypothesis: true difference in means is not equal to 0
## 95 percent confidence interval:
##  29.14294 35.74945
## sample estimates:
## mean of x mean of y
## 33.390201  0.944007
```

Fig. 7 T-test on null hypothesis-1

```
##
##  Welch Two Sample t-test
##
## data:  Spent and Approved_Conversion
## t = 19.609, df = 1142.9, p-value < 2.2e-16
## alternative hypothesis: true difference in means is not equal to 0
## 95 percent confidence interval:
##  45.37197 55.46133
## sample estimates:
## mean of x mean of y
## 51.360656  0.944007
```

Fig. 8 T-test on null hypothesis-2

5.3.3 Hypothesis 3

Null hypothesis: *Frequency of advertisement occurring did not affect the approved conversion.*
Alternate Hypothesis: *Frequency of advertisement occurring affects the approved hypothesis directly.*
Correlation of Impressions to Approved Conversion = 0.6842485
Performing T-test on the Null Hypothesis,

Therefore, null hypothesis is rejected, sales increase with the frequency of advertisements (Fig. 9).

```
##
##  Welch Two Sample t-test
##
## data:  Impressions and Approved_Conversion
## t = 20.185, df = 1142, p-value < 2.2e-16
## alternative hypothesis: true difference in means is not equal to 0
## 95 percent confidence interval:
##  168580.2 204882.2
## sample estimates:
##    mean of x    mean of y
## 1.867321e+05 9.440070e-01
```

Fig. 9 T-test on null hypothesis-3

5.4 Visualizing the Sales and Marketing Data

In order to visually describe the sales and marketing data, the following data visualization techniques have been used,

5.4.1 Boxplots

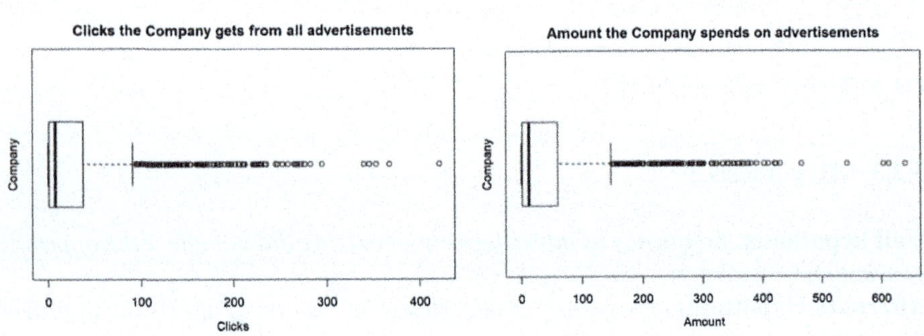

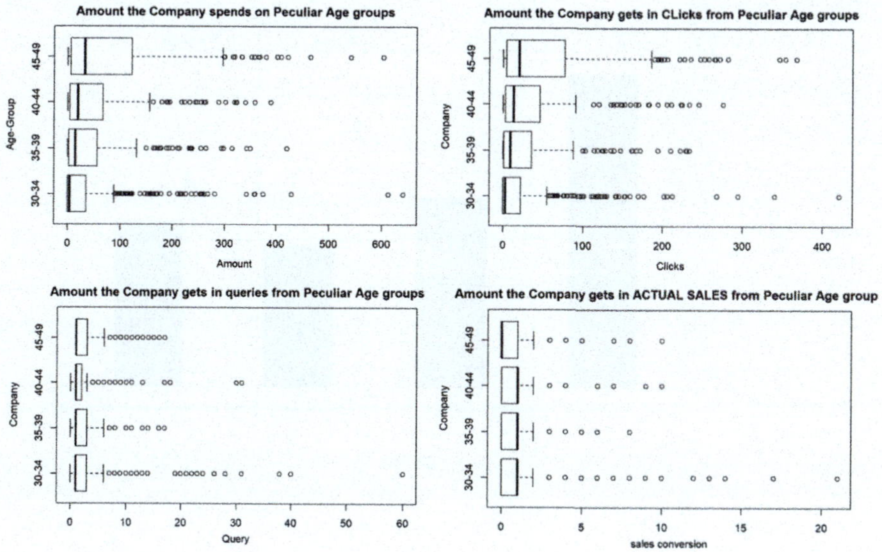

5.4.2 Histograms

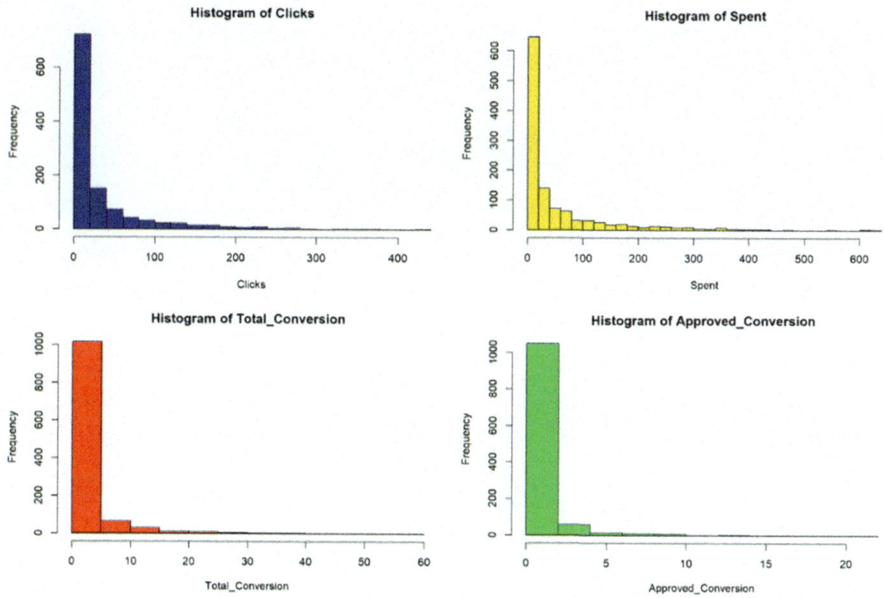

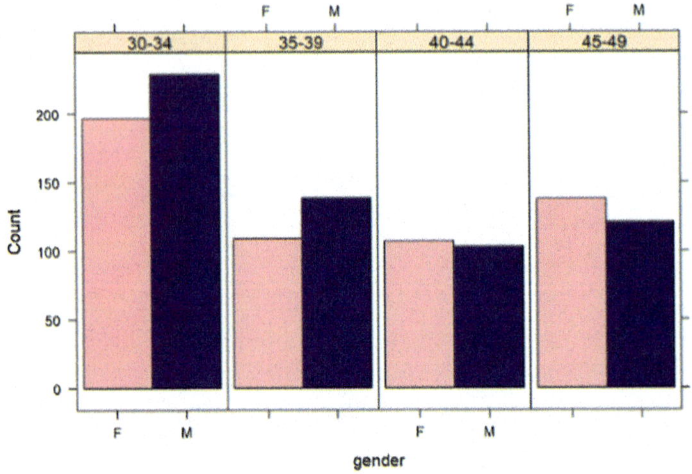

5.4.3 Scatter Plots

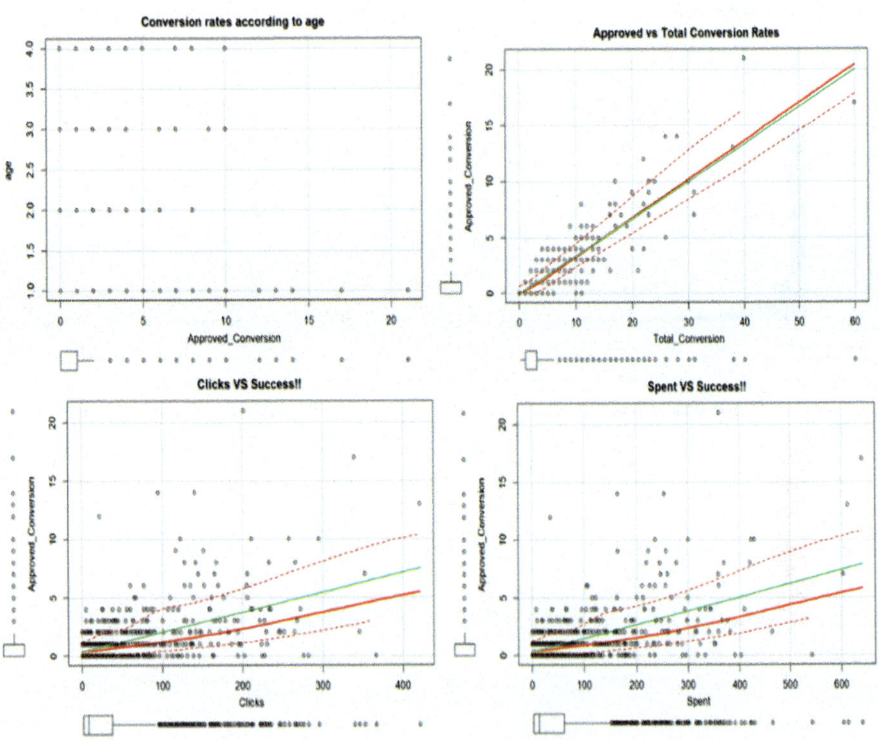

6 Issues, Controversies, and Problems

6.1 The Issues in Retrieval

Content and format of the data presented must be adapted to the stage in the decision-making process and in synchronization with the resultant state of the user [4]. For an instance, if the user has already made a premature decision, without the aid of data, information encapsulating that decision must be presented very tactfully to cover the tendency to disregard such information (even then the user may not accept the information).

A similar situation exists in the provision of data for a marketing research or online business expansion project [5]. The type of data varies as the project travels from the inception and planning phase, which requires overall orientation, to the execution or working phase, which requires techniques on data collection strategies, which could be very dynamic with the passage of time.

6.2 Issues in Getting Access to the Marketing Data for Visualization

There are issues with the security and privacy of data during its retrieval:

- **Data access vulnerability**: *the analyst has access to the customer's personal data*
- **Data breach vulnerability**: *the company suffers a data breach*
- **Spillover vulnerability**: *a company's close opponent suffers a data breach*
- **Data manifest vulnerability**: *a data breach allows customer's information to be misused, maybe for identity theft.*

6.3 Controversies Where Sales Data Was Gathered Illegitimately

To gather sales data, there can be huge controversies to legitimize the data acquisition and prevent any sort of data mismanagement in case vital information of the customer is also taken into account.

Company should have cautioned the customer of his/her information usage for data analysis and mining purposes and customers should be fully aware about the type of data they would be willing to share. Controversies can also arise on targeting every search query of the customer to know his/her interests, particularly private and extremely personal information. The company's data security can also be compromised and this poses a belligerent threat to the stakeholders of the entire company.

7 Problems in Prediction and Visualization

Major problems in Prediction and Visualization hover around the following stages:

1. **The limitations of algorithms**—This is the biggest potential problem and also the most complicated. Any algorithm used is based on human inputs, and human work can be full of errors. For example, a user designing an algorithm may point out different pieces of data that are very important to consider and delete other pieces fully, also other situations, where there are data outliers or unique situations that demand an alternative approach, pose a problem for the already defined methodology.
2. **Over-reliance on Visualizations**—This is more of a problem with consumers and non-developers than it is with analysts, but it brings down the potential impact of visualization in general. When stakeholders begin relying on visuals to comprehend data, they very much begin to over-rely on it. For an instance, they may take the inferences as absolute and pure truth and never delve deeper into the datasets in use for producing those visuals. The conclusions that can be drawn from this may be generally applicable, but they will not tell us everything about the audiences or campaigns.
3. **Constraints due to Advancements**—There exist a lot of visualization software to aid in better analysis of complex datasets via charts, and illustrations, and data visualization is too important to get extinct. The advancement does cause us to be on a fast-track to the betterment of visualization techniques, and there is no real turning back at this point. Some of the very effects would be companies racing to develop visualization products and consumers only looking for products that offer visualization. These effects may fertilize the user's over-reliance on visualization and multiply the limitations of human errors in algorithm development.

8 Solutions

Solutions must focus on mitigating threats to the security and to retain the trust of the customers and where possible to convert the liability to purchase. Routinely adding patches that track and eliminate known and unknown vulnerabilities can be one way of addressing this issue. Working with multiple security providers and applying the principle of least privilege can also help in curbing data breaches and internal threats. Although a system of multiple backups is convenient and may reduce costly cleanups in the event of a breach/data attack, companies must ensure that sensitive information is not unnecessarily moved around in various devices.

Another area that needs to be addressed is data overload. Too much content can lead to over-visualization and may hinder the predictive ability of the same. In the same way, over-simplification of data may leave the analyzer with very little information to work with and may lead to an incomplete visualization and therefore incorrect interpretations. A related problem is over-reliance on visualization. A simple way to overcome this problem is to use the "treasure map technique." This requires one to use physical components as a parallel to logical mental components. An alternate technique is to create a receptive visualization which requires a multimodal model of presentation [6].

A critical area of concern is the human limitations of algorithms. Human inputs are subject to errors and this may cost the company critical information. Apart from the human error factor, one must remember that algorithms are analytical tools that perform operations based on past representations of data. They may predict quantitative changes in the future but are inherently incapable of visualizing qualitative changes in the future.

Hou and Sourina and many others in the field have applied a very creative solution to the problem in visualization [3]. That is the use of haptics in accordance with human physiological characteristics in analyzing datasets. Representation and interaction are two major components in visualization process. Haptics, though not adapted to memorization and recall in terms of interaction.

9 Results

Based on the visualization plots, it can sufficiently be concluded that, the hypothesis that approved conversion (final consumption) depends directly on the:

1. Number of clicks received by the social media advertisement
2. Frequency of the advertisement being displayed in front of users
3. Interest specified by users
4. Amount of queries received by the companies
5. Amount of money spent by the companies toward the advertisement campaign.

This result demonstrates that the social media advertising is a good investment, as long as the impression is made on the correct interest/niche of audience. Thus, the Model 1 is a more efficient revenue generation model.

Model 1 fits better than Model 2 due to lesser AIC value.

Best Fit: $y = Approved\_Conversion \sim Impressions + Spent + interest + Clicks + Total\_Conversion$.

10 Future Research Directions

Future research can be either be done to improve the predictive analysis techniques, taking more variables into account that could affect the sales of the company or to enhance the type of visualization techniques that are implemented [7]. Some of the visualization techniques that can be implemented are:

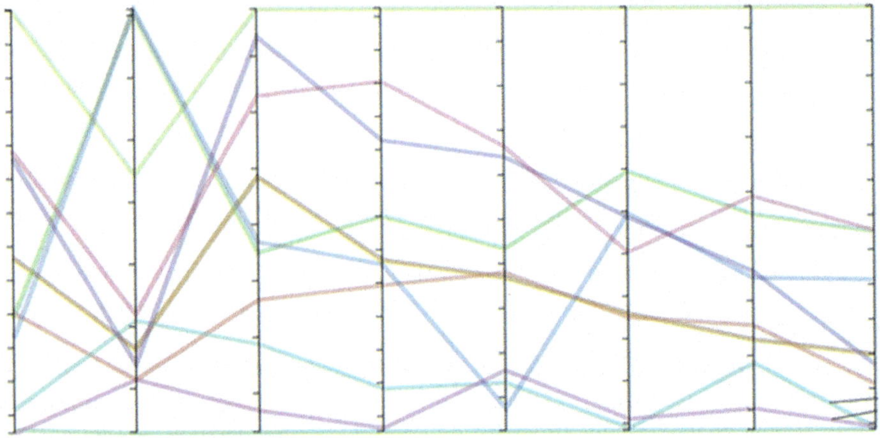

Fig. 10 Parallel coordinate visualization plot

10.1 Parallel Coordinates

To reveal how groups of variables exhibit similar or different profiles across many quantitative variables. Parallel coordinates visualizations are among the best visualization types for large-scale and dynamic data (Fig. 10).

10.2 Alluvial Diagrams

To show how various groups are in relation with one another or are different from one another across many variables. These diagrams are especially useful for literal and temporal flows: of money, goods, and time (Fig. 11).

10.3 Circle Packing

Circle packing diagrams show groups as tightly organized circles and are often used to display hierarchies where smaller groups are either colored similarly to others in the same category or nested within larger groups (Fig. 12).

Fig. 11 Alluvial visualization plot

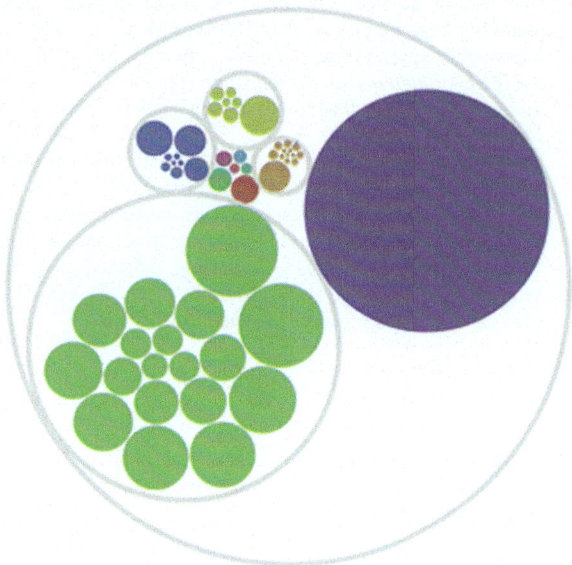

Fig. 12 Circle packing visualization plot

11 Conclusion

This paper was motivated by the need for research that could improve the understanding of how social media advertising influences the sales of products in the online shopping industry. The results confirm the hypotheses that the product consumption by end users is dependent upon or is a function of the interaction of the variables, namely the number of clicks, the company's advertisement budget, the focus on the areas of interest specifically expressed by the customer, and analysis of queries that the company receives. Data visualization is an extraordinary tool that if used effectively, can benefit the company and the end user enormously. It is actually a virtual interface that brings out the patterns of consumer activity like a *bas relief.* Several studies are being conducted to optimize the visualization methods and the results of these studies are confirming the usefulness and effectiveness of data visualization as a tool. Data visualization is cost-effective and efficient and holds tremendous potential as a consumer-company liaison tool.

References

1. Ringel, D. M., & Skiera, B. (2018). 19. visualizing asymmetric competitive market structure in large markets1. *Handbook of marketing analytics: methods and applications in marketing management, public policy, and litigation support* (p. 431).
2. Gründemann, T., & Burghardt, D. (2018). Classifying and visualizing the social facet of location-based social network data. In *VGI Geovisual Analytics Workshop, colocated with BDVA 2018.*
3. Hou, X., & Sourina, O. (2016). Real-time adaptive prediction method for smooth haptic rendering. arXiv preprint arXiv:1603.06674.
4. Christopher, G. H. (1996). Choosing effective colours for data visualization. In *Proceedings of the IEEE Conference on Visualization* (pp. 263–270).
5. Tukey, J. W. (1962). The future of data analysis. *The Annals of Mathematical Statistics, 33*(1), 1–67.
6. Bangay, S. Visview: A system for the visualization of multi-dimensional data. In *Visual Data Exploration and Analysis V.* (TA 1505 Pse 3298).
7. Foley, J., & Ribarsky, B. Next-generation data visualization tools. In L. Rosenblum, R. A. Earnshaw, J. Encarnacao, H. Hagen, A. Kaufman, S. Klimenko, G. Nielson, F. Post, & D. Thalmann (Eds.), *Scientific visualization, advances and challenges.* (T385 Sci).
8. Wong, P. C., Bergeron, R. D., Nielson, G. M., Hagen, H., Muller, H. *Scientific visualization, over view, methodologies, techniques.* (Q175 Nie.).

Applications of Visualization Techniques

A Case Study on Political Event Detection

Deepak Kochhar, S. P. Meenakshi and Satakshi Dubey

> A picture can say a thousand words, but a graph can say a lot more than that.

Abstract We are living in a data-driven society, and everything you see and do is data. Over 2.5 quintillion bytes of data are created every single day, and it is only going to grow from there. By 2020, it is estimated that 1.7 MB of data will be created every second for every person on earth. In such a scenario, it is almost impossible, labor-intensive, and time-consuming to mine data in traditional ways to bring out insights from it. Data holds the key to smart decision making, and almost every disruptive technology today highly depends on it, but if we do not devise an efficient way to dive into data and map learnings from it, then this potential resource is of no use. To all the doubts and required advancements, data visualization is the answer. Data visualization provides a way to represent quantitative data in a graphical manner. It holds the potential to transform any kind of data into visuals; something which is easier to perceive and process by the human mind. Through data visualization, we can map years of messy bulky data into expressive visualizations to discover new trends and unknown facts. The visualization methods vary from the trivial line charts to the bar, column, pie, heat maps and what not. Good data visualizations are created when intelligent communication, data science skills, and impressive design techniques collide. It offers key insights into complicated datasets in ways that are meaningful and intuitive. It is a highly versatile field of research and development and finds enormous applications in the world of business intelligence that drives industries, in education and learning space to better communicate ideas, in geospatial studies, social network analysis, prediction analysis, and an insanely huge number of other fields. New ways to incorporate data visualization into work evolve every

D. Kochhar (✉) · S. P. Meenakshi
School of Computer Science and Engineering, VIT Vellore, Vellore, India

S. Dubey
Department of Computer Science, JEC Jabalpur, Jabalpur, India

S. M. Anouncia et al. (eds.), *Data Visualization*,
https://doi.org/10.1007/978-981-15-2282-6_6

day. The rapid development of data visualization tools and technologies has enabled harnessing of data and transforming it in a way that turns it into information. In this chapter, we answer the question of why we should use data visualization along with discussing associated technologies and impressive hands-on applications to support our reasons.

Keywords Data visualization · Storytelling · Word clouds · Heat maps · Bar chart · Line chart · Event detection · Electoral campaign · Facebook micro-ads

1 Why Do We Use Data Visualization?

While there can be many reasons to do so, we have listed the top three reasons below:

1. **Meaningful Storytelling**—We are already overwhelmed by the huge volume of data we have to deal with each day, so poring over large spreadsheets is not an ideal option. Data visualization helps to condense and analyze data and turn it into a meaningful concept. Professionals and organizations can leverage the power of visualization techniques to derive an insightful interpretation.
2. **Better decision making**—Data is the new oil, and it fuels the critical decision-making process across organizations. This process is made possible on account of credible insights obtained from data via spatial and temporal analysis. Data visualization helps in communicating those insights better and more vividly and thus helps in devising strategies, make better decisions and better planning.
3. **Data Literacy**—According to Wikipedia "*Data literacy is the ability to read, understand, create and communicate data as information. Much like literacy as a general concept, data literacy focuses on the competencies involved in working with data.*" Data literacy is often quoted as a critical skill for the twenty-first century. A new breed of data-intensive organizations is sprouting all around us that are leveraging data literacy to upskill their workforce. Data visualization is what impacts data literacy the most, and thus, it is a critical component.

2 Applications of Data Visualization in the Real World

2.1 Data Visualization for In-House Communication and Client Reporting—Business Intelligence

Data visualization offers a technology-driven process for analyzing corporate data and presenting actionable information in the form of impressive visuals to help executives, managers, and other corporate end users make informed business decisions. The data visualization technologies can map historical, current, and predictive views of business operations to devise smart business strategies. It can be used to model consumer/client behavior to better understand their requirements and levels of satisfaction and to identify groups of clients with similar or specific needs to serve them accordingly. It can also be used to effectively discover the target segments. A data-oriented mindset in the corporate space is a competitive advantage. Leading organizations around the world have realized that business data and content are not to be managed separately from the rest of the information management, and they continually make use of visualization technologies to work on data and equip their business with intelligence. The omnipresent data is the raw material for businesses to discover trends and information. It enables one to perform sophisticated analyses and glean insights even without a strong technical background since the visual context is always easy to understand and much more expressive the textual counterpart.

2.2 *Marketing Content and Data Visualization*

Data visualization and content marketing are often quoted as a match made in heaven. Data visualization is a critical skill for modern marketers. The information flow is insanely huge, and so, it mainly serves two purposes in the marketing space. The first and very obvious is it serves a prosaic purpose. It easily gets people's attention. If users are confronted with the messy boring data talking about a product, it is highly certain that you will gain nothing, whereas if they are presented with more human-friendly and attractive visuals, there are high chances that you are going to ace it. The second purpose it serves in the marketing sphere is that it solves the problem of relaying complex information. Visualization is a way of making sense of all the data, ideas, and information. Data visualization can not only deliver marketing insights but can itself be a content to fuel smart decision making. Data can generate interesting lessons to share. The following visual illustrates this stunning application. The visuals enable better connection and communication between the consumers and the market and help to discover the target audience.

2.3 *Data Visualization for Text Mining—Semantic Technology*

With the increasing intricacy and size of data, the transformation from streaming knowledge into actionable information becomes more and more challenging and requires a synthesis of computational approaches. Text mining and analytics have progressed with techniques such as entity extraction and characterization, topic and opinion modeling, and sentiment and emotion analysis, but text visualization has not advanced much since the tag cloud. Recent advances in the field of data visualization have opened a new scope for data visualization applications in the text mining space.

The textual data from social media and other relevant sources carries key semantics to discover relationships. Deep semantic analysis enables a complete understanding of the text to exploit the data, and further transforming this knowledge to visual context reveals hidden relationships and captures even the weakest signals present in information. Data visualization can capture trends in society by mining and presenting popular #hashtags. It can also be used to judge the general sentiment of the public toward the governmental activities by visual modeling of their related online activities. A person's behavioral analysis can be performed by mapping his speech (most spoken words) to a visual to identify his intent. The integration of the fields of text analytics and data visualization is happening rapidly and shows promising results.

2.4 Collaborative Visual Analysis—Exploring and Making Sense of Data with Others

Data visualization and visual analytics are promising technologies for tackling complex and dynamic datasets. It leverages the visual systems to support the sensemaking process. Sensemaking is a social process, but most of the studies assume a single-user perspective as is evident from research studies conducted till date. To fully support sensemaking, interactive visualization should also support social interaction. The collaborative visual analysis is a solution that can enable research teams to reach greater insight into interdisciplinary projects. It can enhance decision making by facilitating social interpretation and the enablement of distributed exploration which allows users to pool in their efforts and findings as they mine and analyze the data.

3 Data Visualization Techniques

1. Word Clouds

Word clouds are used to get an idea of how often a word is used, and it shows what is emphasized in your text. The words are of different sizes in a generated word cloud, and a larger size implies a higher frequency. Instead of poring over text to develop any lead, we can generate word clouds to reveal what is essential. Word clouds are fast to develop and engaging. This visual representation of text data has an engaging impact and generates an observer's interest. The word cloud thus helps to provide an overall meaning to the data which does not happen when you just read the same data. Word clouds help in sharing back results from research in a way that does not require an understanding of the technicalities (Fig. 1).

This technique can be applied to a lot of use cases. For business, it can be used to identify your customer's grievance; the largest word generated out of customer feedback data would be the top concern you need to care about. In our case of political

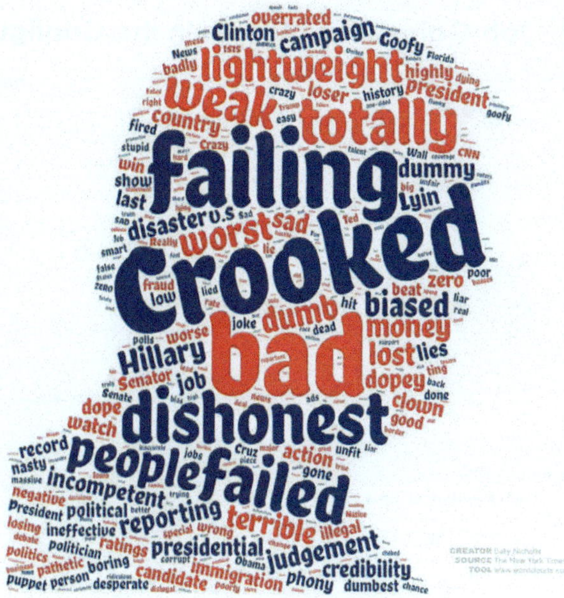

Fig. 1 Word clouds. *Source* The New York Times

event detection, we have collected the speech data of the top two US electoral candidates and generated word clouds out of them. Those word clouds (discussed later on) provide an insight of the most used words by each candidate which subsequently reveals their political ideology, policies or the issues they were most concerned about. Interestingly and quite obviously, the word cloud also reveals some of the famous hashtags that went viral on social platforms then.

2. **Symbol maps**

Maps with symbols on them make up a symbol map. It helps in building a visual to put data in relation to its location. Symbol maps proportionally scale the size of symbols used to the data value for that specific location. For example, if we draw a symbol map of Indian cities based on population, the symbol for Mumbai will be bigger than that of Bengaluru. The symbols differ in size, which makes them easy to compare. Symbol maps are flexible as they allow the use of both numerical (population) and categorical data (low risk, high risk). Due to the geographic element, symbol maps are quite an appealing data visualization tool. It can accommodate a single as well as multivariable data. Congestion and overlapping remain a problem for them in case of high variations and close data locations (Fig. 2).

3. **Connectivity charts**

Connectivity charts depict the links between phenomena, events, persons, etc. The connectivity chart is a kind of network graph which reveals some of the most interesting insights from a variety of data such as detecting communities with a similar opinion and recognizing the most influential person in a network. It provides a method to

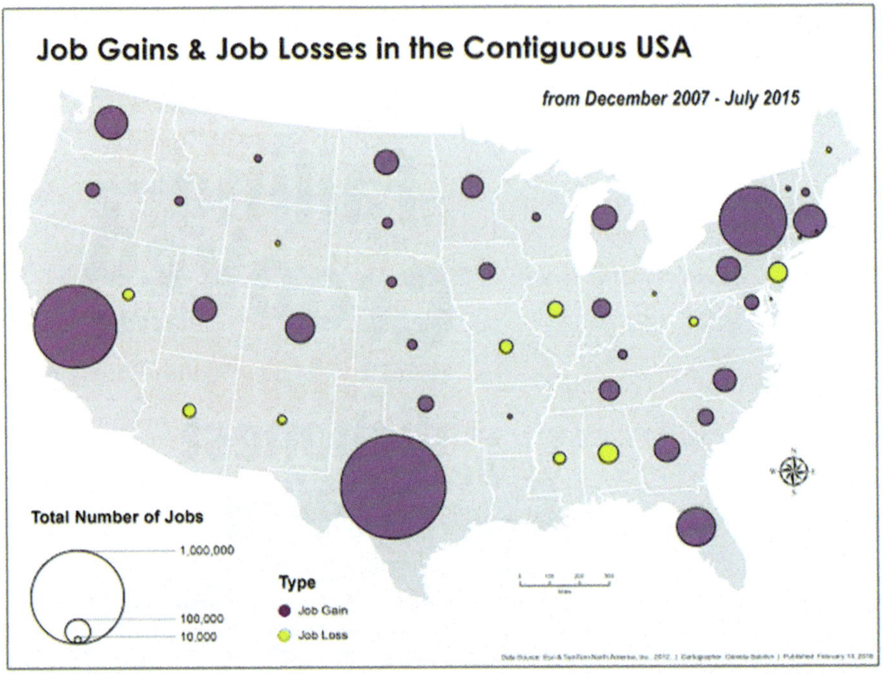

Fig. 2 Symbol maps. *Source* Daniela Sabillon

investigate social structures. These are often visualized through sociograms in which the nodes are represented as vertices and ties as edges. The visualization provides a means of qualitatively assessing networks by varying the visual representation of their nodes and edges to reflect attributes of interest. Although it can be applied to various fields including airline network, banking network, supply chain network to name a few, its usage has gained much popularity in social network analysis. Social network analysis has emerged as a key technique in modern sociology and connectivity charts provide a quantitative and qualitative analysis of that (Fig. 3).

4. **Line Charts**

Line charts allow looking at the behavior of one or several variables over time and identifying the trends. It can show how variables such as sales and profit, etc., change with respect to some other variable such as time. The longitudinal aptitude of a line graph is its most important aspect. Using line charts, we can easily compare two or more factors such as the performance of employee X and Y for the last month. Line charts are easy to use and can also be combined with other kinds of charts as per requirements. In our case study, which is discussed later on, we have used line charts to compare the Facebook posts likes, shares, and comments of two electoral candidates over a period of ten months. It simplistically and beautifully reveals which candidate was ahead, when, and by how much and thus helps in forming decisions out of that (Fig. 4).

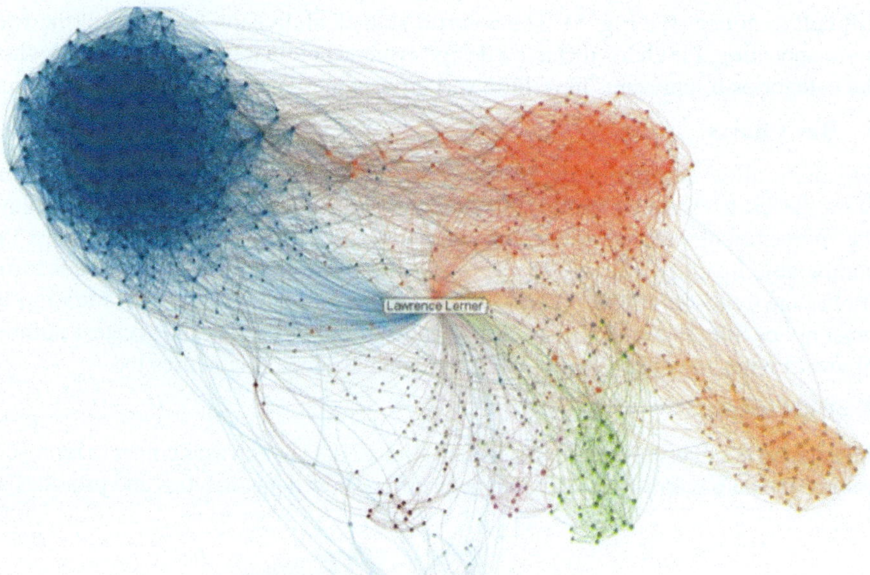

Fig. 3 Community connectivity chart. *Source* Google Images

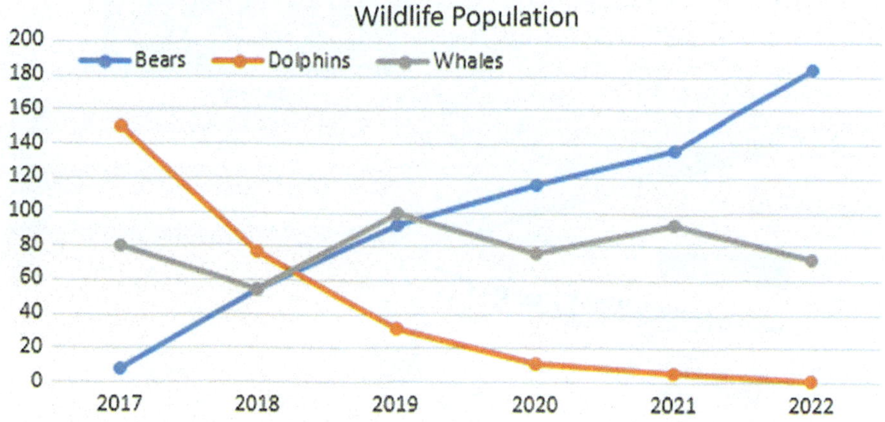

Fig. 4 Line chart. *Source* Google Images

5. **Pie Charts**

Pie charts show the components of the whole. A pie chart is a circular statistical graphic, which is divided into slices to illustrate numerical proportion (for example, the share of voters or population of cities). In a pie chart, the quantity of variable is proportional to arc length. It supports a single variable. Pie charts are useful for displaying data that are classified into nominal or ordinal categories. It is an ideal practice not to include more than six categories in a pie chart, otherwise, it becomes difficult to notice (see Fig. 5). The colorful statistical division becomes all the way more appealing. Pie charts do have a design issue, wherein it becomes cluttered when the categories it represents have little variations which make interpretation difficult.

6. **Bar Charts**

Bar charts allow comparing the values of different variables. The bar charts are one of the easiest to create which also makes it one of the most misused charts. It displays the frequency of items in each category, and the length of each bar represents the comparable quantity. It can effectively summarize a large data set in visual form. It clarifies trends better than tables do. The tallest and the shortest bars tend to be most noticeable. Figure 6 shows the bar chart of directors with the greatest number of award wins, the effective use of legends enables depicting genres too.

7. **Heat maps**

A heat map is a two-dimensional representation of data which uses colors to represent values. It uses a warm-to-cool color spectrum to show data analytics and provides an

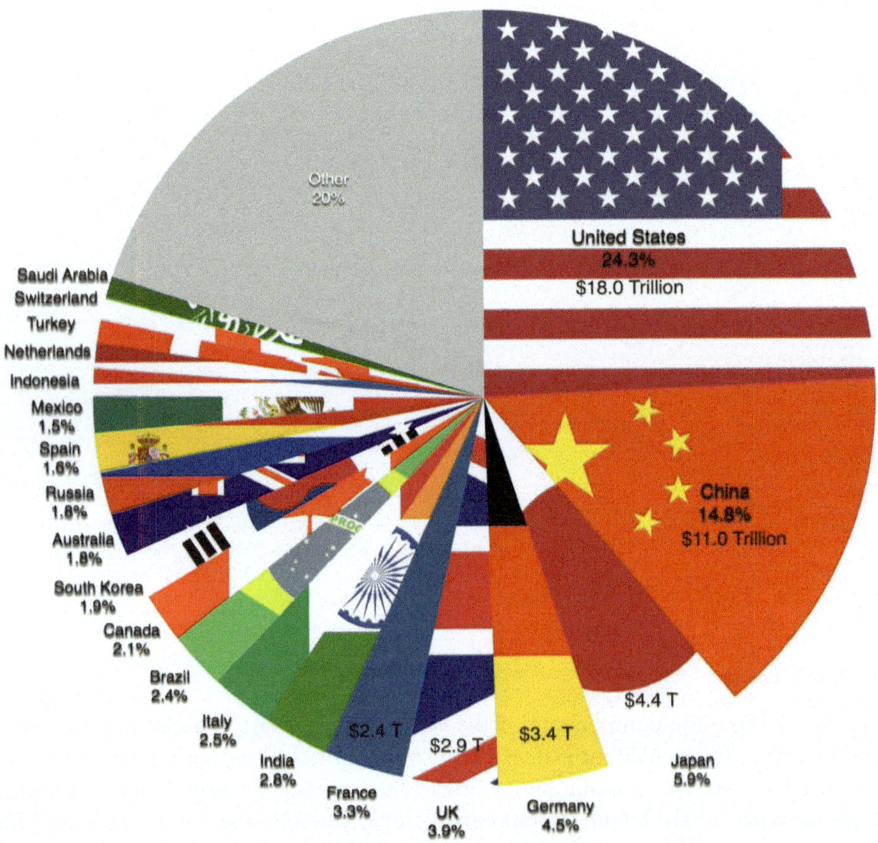

Fig. 5 Pie chart—Global GDP (2015). *Source* Wikimedia Commons

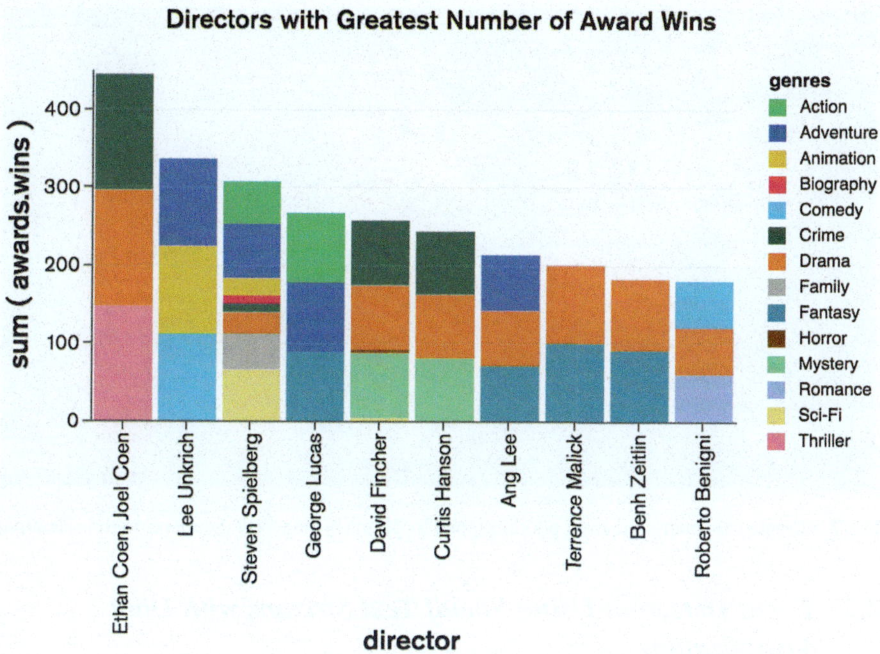

Fig. 6 Bar chart. *Source* MongoDB Documentation

immediate visual summary of information, for example, during an election, a geographical heat map can inform the readers which states each candidate has won. We have used a similar kind of election heat map to show which US electoral candidate has won from a given location which subsequently depicts the candidate majority. There are temperature heat maps too which show the temperature values across locations, the more intense the color, the higher is the temperature (red for a warmer climate and blue for a colder climate) (Fig. 7).

4 Case Study

In this section, we are including a case study about how data visualization can fetch and reveal some of the very interesting insights from the US presidential elections of 2016. This will help the readers to better understand and connect the practicality of the concepts to the real world.

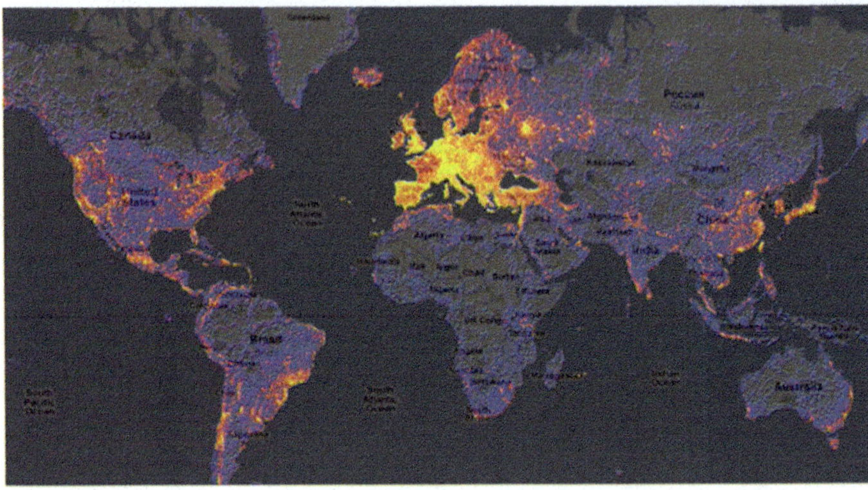

Fig. 7 Google heat map—the most photographed locations in the world. *Source* Google's Heatmap

4.1 Event Detection Using Social Text Streams with Data Visualization

A large quantity of structured or semi-structured textual media exhaustively contains information about events, such as electoral trends, crimes, environmental problems, and consumer trends. Recently, social text stream sources of information like weblogs, message boards, mailing lists have become omnipresent with the rapid evolution of the Internet. Social text stream data is outlined as a group of informal text communication data that arrives over time, and each piece of text in the stream is related with some of the social attributes such as writer, reader, and recipients. Usually, social text stream information arrives over time, and each and every piece of the stream carries part of the semantics (e.g., data regarding real-world events). Social text streams serve as the sensors of the real world. With a colossal quantity of data and numerous varieties of sources in social text data, combined with the rich content such as text, relations, social actors and temporal information, expeditiously organizing and summarizing the embedded semantics has become a very important issue. Most text semantic analysis techniques are principally targeted on formal text stream data. However, the social text stream data is substantially different from formal text stream data:

(1) social text stream data is way more context sensitive and
(2) social text stream data contains rich social connections.

Given the distinguishing features of social text stream data and the rich information embedded in them, we can detect events from them. Social text stream data also tells us a lot about social structures, and it can help to build a social network which can be subsequently mined to extract insights. We are indirectly referring to Facebook

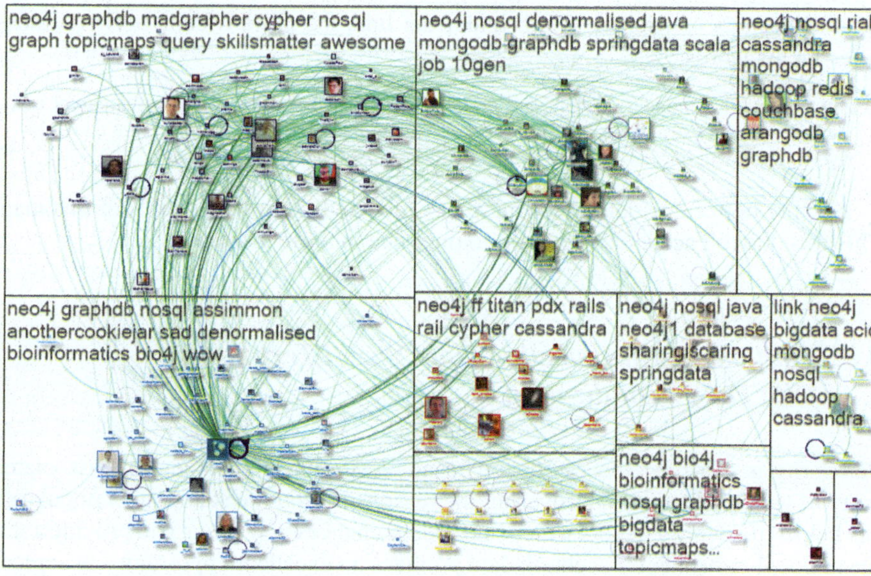

Fig. 8 The graph represents a network of 230 Twitter users whose recent tweets contained "neo4j." There is an edge for each 'reply' and 'mention.' *Source* NodeXL Graph Gallery

social graphs in our work which lets you know who all are your friends, whom you interact with the most, who are your mutual connections, etc. An example of a Twitter social network is shown in Fig. 8.

The social data contains a lot of hidden information. The graph shown in Fig. 9 was made by Paul Butler, a Facebook intern. The map was the result of his work to visualize where people live relative to their Facebook friends. Each line connects

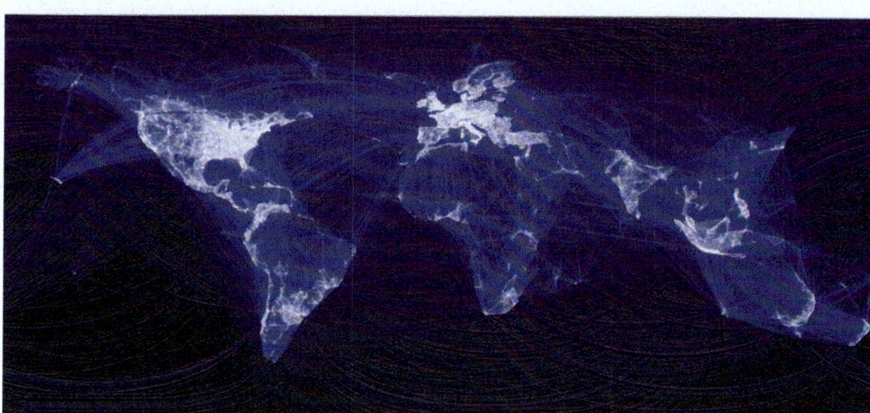

Fig. 9 Facebook connections mapping the world. *Source* Facebook

cities with pairs of friends. The brighter the line, the more friends between those cities. In his words,

> Not only were continents visible, but certain international borders were also apparent as well. What really struck me, though, was knowing that the lines did not represent coasts or rivers or political borders but real human relationships.

Another noticeable detail was missing China and Central Africa from this map. This was due to Facebook's little or no presence in these locations.

4.2 Political Event Detection from Social Text Streams with Data Visualization

Popularity is a critical success factor for an electoral candidate and his associated party to win an election. Discovering the reasons behind the candidate's popularity can help us to identify factors which boosted his chances of winning and provide a stable political movement. Facebook data provides an excellent opportunity to explore public opinions by analyzing a large number of Facebook posts. In this work, we attempt to measure and compare the popularity of two US electoral candidates of 2016 elections in Facebook using a set of three parameters namely likes, shares, and comments. The number of likes on a post indicates how viral the post went, the comments indicate how engaging it became, and finally, shares indicate how much a candidate's idea was shared among his followers. Increase in connections, likes, shares, and comments can be inferred as increasing popularity. The analysis of the development of popularity is a case of political event detection. At last, the candidate who has the most likes, comments, and shares will obviously have higher popularity and will have higher chances of securing maximum votes and winning elections. This is because a lot of people especially in countries like the USA nowadays extensively use social platforms to shape their opinions about candidates and parties, and Facebook being a social media giant plays a crucial role in that. We used data visualization techniques to analyze the Facebook data, and the results were quite interesting (discussed in upcoming sections).

4.3 Problem Statement

Analyze the role of micro-advertising on Facebook in the US Electoral Campaign using techniques of data visualization.

Micro-targeted advertising on Facebook was very effective in persuading undecided voters to support Donald Trump during the 2016 US Presidential Elections. The targeted advertisements were powerful enough to polarize the voters toward Donald Trump whose team spent 44 million dollars on these advertisements. As a result of which, even though the Democrat candidate Hillary Clinton carried fair support among the educated and aware people of the US, Trump managed to secure the support of non-aligned who used Facebook as a major source of news and were not carrying a university or high-school degree.

4.4 *Solution*

We have done a comprehensive study and analysis on the US Presidential Election data using the data visualization and social network analysis techniques which enabled us to more vividly visualize and extract insights into the election behavior. We used the Facebook API along with the RFacebook package of R language to collect data from the official Facebook handles of presidential candidates for almost ten months. We have further used ggplot2 + plotly and shiny to generate insightful visualizations.

4.4.1 Techniques and Tech-Stack Used

1. Facebook API
2. RFacebook package
3. Shiny
4. ggplot2
5. Google sheets
6. Python and R
7. Data visualization and social network analysis techniques.

Step 1—We collected data from the official Facebook pages of both the major candidates for a period of about ten months. We started from January 1, 2016 to the date of the election that is November 8, 2016.
Step 2—We used the **RFacebook package (from the R language)** and Facebook API to mine and collect this data. Due to the limitation of Facebook API, it was not possible to collect all the data in one go, so we collected them month-wise.
Step 3—We were able to collect about 500 posts for Donald Trump and 653 for Hillary Clinton. We analyzed these posts on the basis of the count of likes, comments, and shares these posts gathered in this duration. Figure 10 is an R code snippet to collect data.
Step 4—We tabulated the month-wise statistics of these parameters for both the candidates and then plotted them in a temporal manner for each parameter to draw out a distinct comparison (Table 1 and Graph 1).

Conclusion—Trump's posts dominated almost every time, although there was an interesting jump in September of Hillary's posts it was Trump who ended being on top.

```r
trump <- getPage("donaldtrump", token, n = 100, since='2016/01/01', until='2016/1/31')
trumpa <- getPage("donaldtrump", token, n = 100, since='2016/02/01', until='2016/02/28')
trumpb <- getPage("donaldtrump", token, n = 100, since='2016/03/01', until='2016/03/31')
trumpc <- getPage("donaldtrump", token, n = 100, since='2016/04/01', until='2016/04/30')
trumpd <- getPage("donaldtrump", token, n = 100, since='2016/05/01', until='2016/05/31')
trumpe <- getPage("donaldtrump", token, n = 100, since='2016/06/01', until='2016/06/30')
trumpf <- getPage("donaldtrump", token, n = 100, since='2016/07/01', until='2016/07/31')
trumpg <- getPage("donaldtrump", token, n = 100, since='2016/08/01', until='2016/08/31')
trumph <- getPage("donaldtrump", token, n = 100, since='2016/09/01', until='2016/09/30')
trumpi <- getPage("donaldtrump", token, n = 100, since='2016/10/01', until='2016/10/31')
trumpj <- getPage("donaldtrump", token, n = 100, since='2016/11/01', until='2016/11/09')

clintona <- getPage("hillaryclinton", token, n = 100, since='2016/01/01', until='2016/01/31')
clintonb <- getPage("hillaryclinton", token, n = 100, since='2016/02/01', until='2016/02/28')
clintonc <- getPage("hillaryclinton", token, n = 100, since='2016/03/01', until='2016/03/31')
clintond <- getPage("hillaryclinton", token, n = 100, since='2016/04/01', until='2016/04/30')
clintone <- getPage("hillaryclinton", token, n = 100, since='2016/05/01', until='2016/05/31')
clintonf <- getPage("hillaryclinton", token, n = 100, since='2016/06/01', until='2016/06/30')
clintong <- getPage("hillaryclinton", token, n = 100, since='2016/07/01', until='2016/07/31')
clintonh <- getPage("hillaryclinton", token, n = 100, since='2016/08/01', until='2016/08/31')
clintoni <- getPage("hillaryclinton", token, n = 100, since='2016/09/01', until='2016/09/30')
clintonj <- getPage("hillaryclinton", token, n = 100, since='2016/10/01', until='2016/10/31')
clintonk <- getPage("hillaryclinton", token, n = 120, since='2016/11/01', until='2016/11/09')
```

Fig. 10 R code snippet to acquire data

Table 1 Month-wise statistics for likes gathered

Months	Trump	Hillary
Jan	4,204,033	1,069,163
Feb	3,935,246	1,244,663
Mar	3,485,663	1,347,153
Apr	3,175,620	953,562
May	3,661,899	2,350,750
Jun	4,062,128	1,273,244
Jul	3,023,891	1,814,300
Aug	3,592,787	2,193,618
Sept	4,088,234	4,788,450
Oct	4,999,103	2,760,880

Note Since the figures for the month of November were comparatively small due to less number of days (9 days), we have merged those figures with the October month for a better visual

Similarly, we tabulated and plotted the values for the other two parameters (comments and shares), and the results that came out were quite interesting (Table 2 and Graph 2).

Conclusion—Trump's posts were able to involve more users as a result of which they gathered way more involvement as indicated by the huge numbers of comments. The comments may be both in favor or against, but the mere involvement puts Trump on top here again.

See Table 3 and Graph 3.

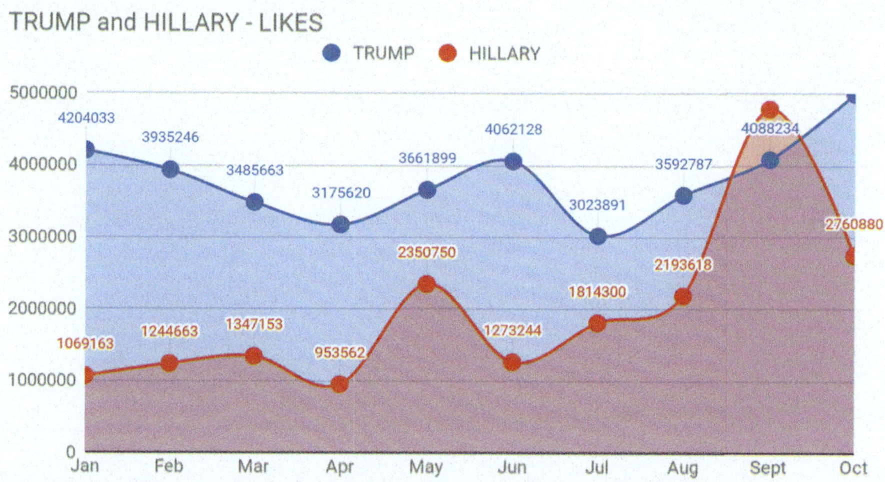

Graph 1 Month-wise comparison for likes gathered by Trump and Hillary's posts

Table 2 Month-wise statistics for comments gathered

Months	Trump	Hillary
Jan	378,863	121,049
Feb	324,726	225,467
Mar	308,072	230,917
Apr	293,586	146,089
May	362,272	590,964
Jun	501,500	441,187
Jul	489,230	412,978
Aug	3,467,969	467,886
Sept	2,917,608	955,316
Oct	4,296,949	767,281

Conclusion—In September, both candidates were in close competition, but after that, interestingly, Trump rose by leaps and bounds, while Hillary's figures sunk. The shares signify whose ideology was most communicated among the public.

But according to HuffPost Pollster, which conducted more than 300 polls across the USA for figuring out the public opinion, the model of the polls suggested Clinton was very likely leading. (In >99% of simulations, Clinton lead Trump) (Graph 4).

The result of US Presidential Elections 2016—Donald Trump sworn in as 45th President of the USA (Fig. 11).

The study and analysis show that Facebook ads helped Trump to turn a large segment of voters who were the primary stakeholders and primarily followed social platforms for shaping individual opinions about elections. In times like this and countries like America, social media play a huge role in shaping public opinion.

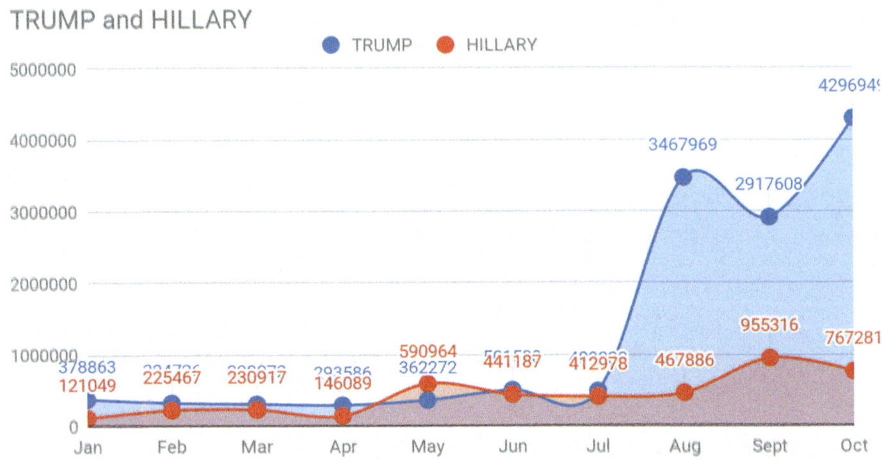

Graph 2 Month-wise comparison for comments gathered by Trump and Hillary's posts

Table 3 Month-wise statistics for shares gathered

Months	Trump	Hillary
Jan	825,927	204,727
Feb	674,157	318,082
Mar	542,492	384,158
Apr	735,834	297,097
May	862,384	942,934
Jun	1,972,514	407,558
Jul	881,047	813,235
Aug	904,575	1,117,368
Sept	1,190,186	1,205,478
Oct	2,148,793	546,305

Data visualization helps us beautifully to understand and dive deep into these months of enormous data to catch interesting insights that reveal some of the lesser-known event facts. Data visualization techniques are versatile and can be applied to a range of domains to discover information.

Step 5—We actually took a step further and collected the top three speeches both candidates made that gathered popular involvement and generated word clouds to discover words which the candidates most emphasized on. We gathered speech data for the following speeches.

Trump's Election Speeches

1. **Charlotte, North Carolina rally** (Fig. 12)
2. **Phoenix, Arizona** (Fig. 13)

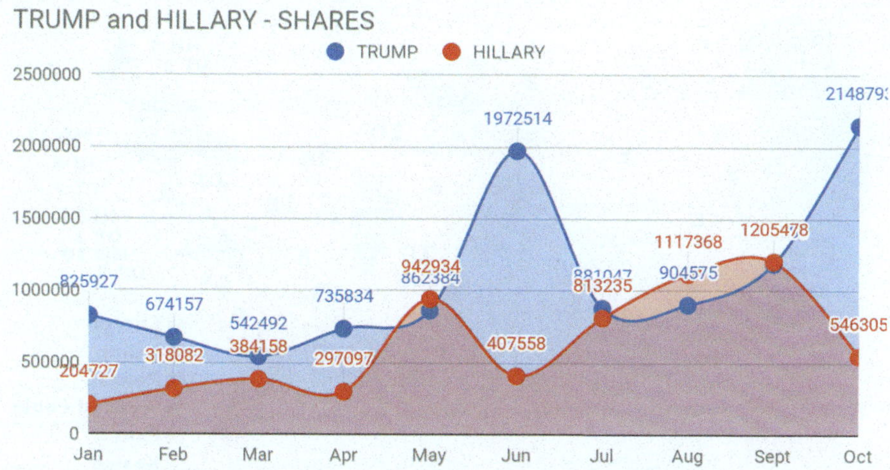

Graph 3 Month-wise comparison for shares gathered by Trump and Hillary's posts

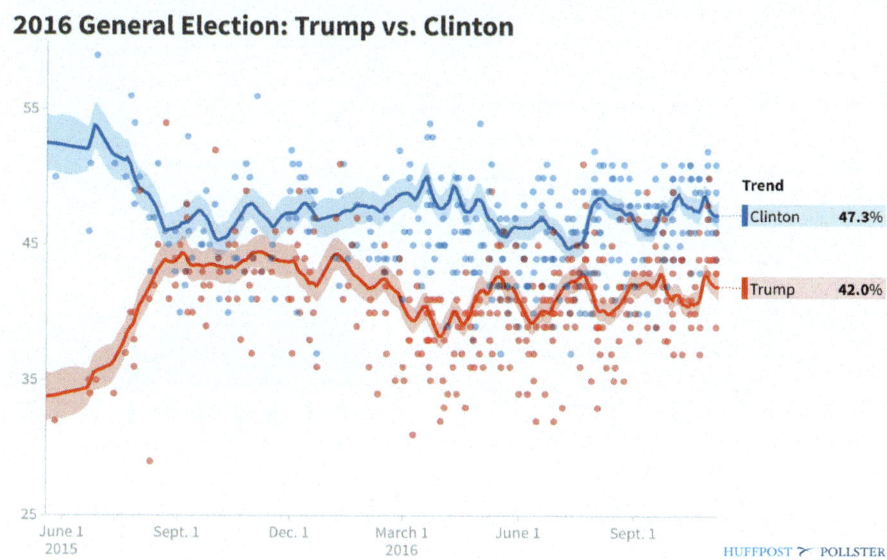

Graph 4 HuffPost Pollster's public polls graph which clearly showed the general public opinion favored Clinton from the very beginning to the end. This graph can be found on the HuffPost Pollster website. *Source* HuffPost Pollster

3. **Donald Trump NYC speech on stakes of the election** (Fig. 14)

Analysis—Trump's speech word clouds reveal that he focused strongly on two things—policies and issues such as immigration and jobs. and criticizing his rival

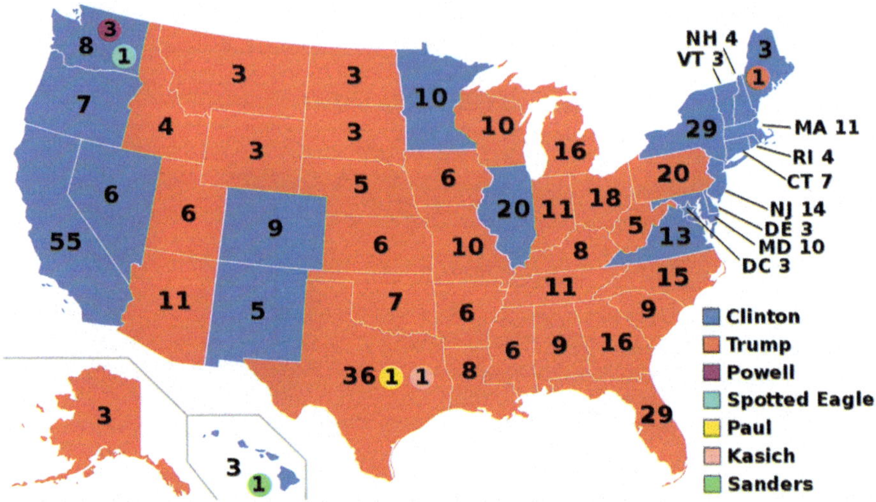

Fig. 11 Presidential election results map (data visualization application). *Source* Wikipedia

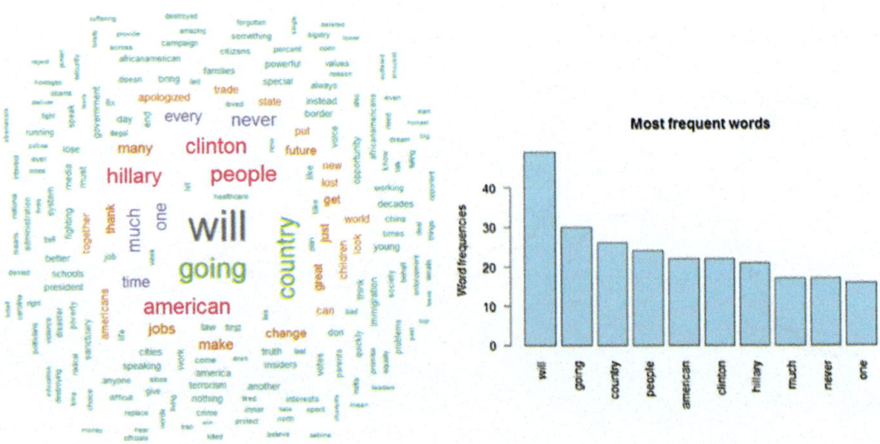

Fig. 12 Word cloud and bar chart for Trump's word usage in North Carolina rally speech

Hillary and her decisions. His **#makeamericagreatagain** wave also became tremendously popular and drew people toward him. He stressed words like **country, people, America** the most which connected him closely with his audience and thus emphasized his views more strongly.

Clinton's Election Speeches

1. **Ohio State University** (Fig. 15)
2. **Nevada** (Fig. 16)
3. **Democratic Convention**(Fig. 17)

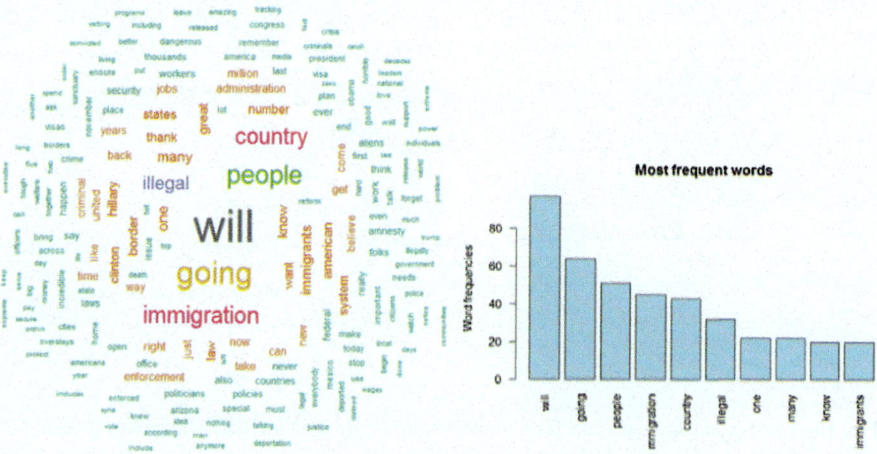

Fig. 13 Word cloud and bar chart for Trump's word usage in Arizona speech

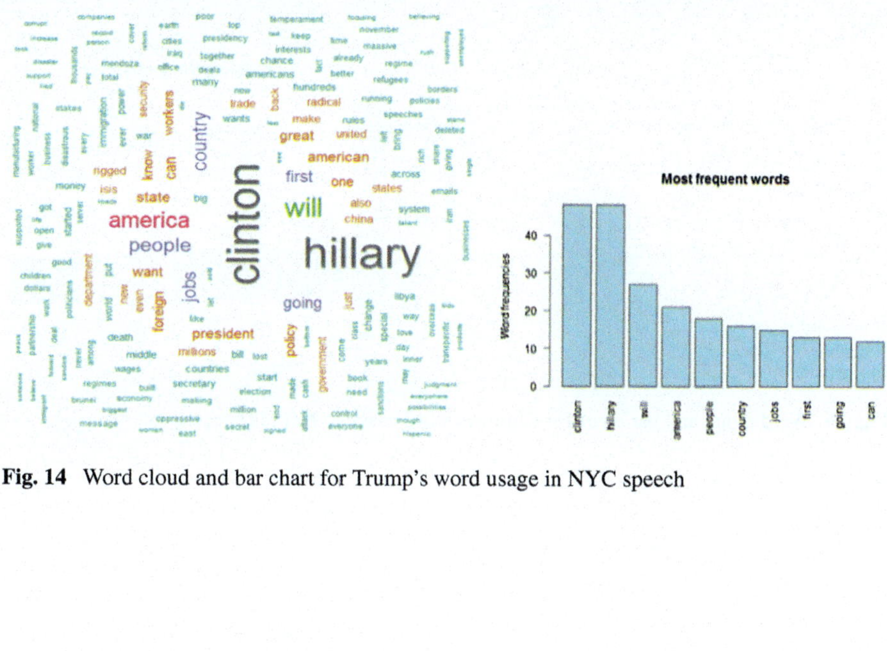

Fig. 14 Word cloud and bar chart for Trump's word usage in NYC speech

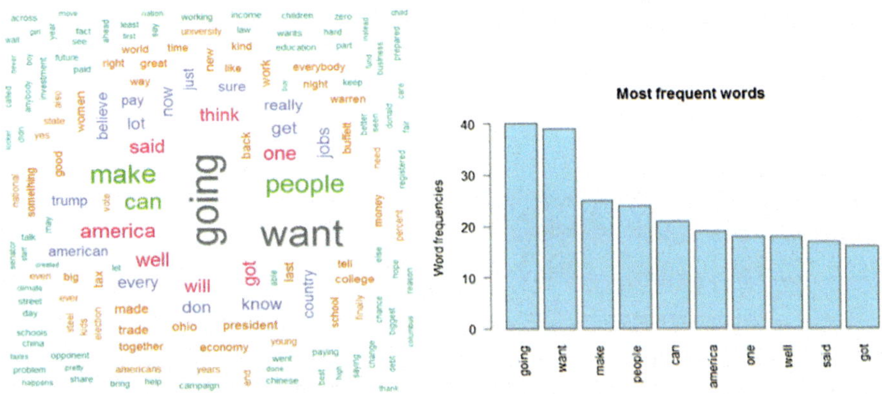

Fig. 15 Word cloud and bar chart for Clinton's word usage in Ohio State University speech

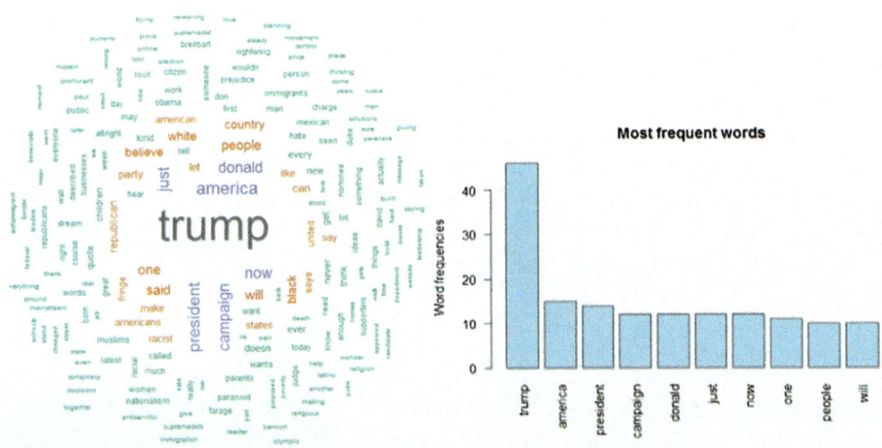

Fig. 16 Word cloud and bar chart for Clinton's word usage in Nevada speech

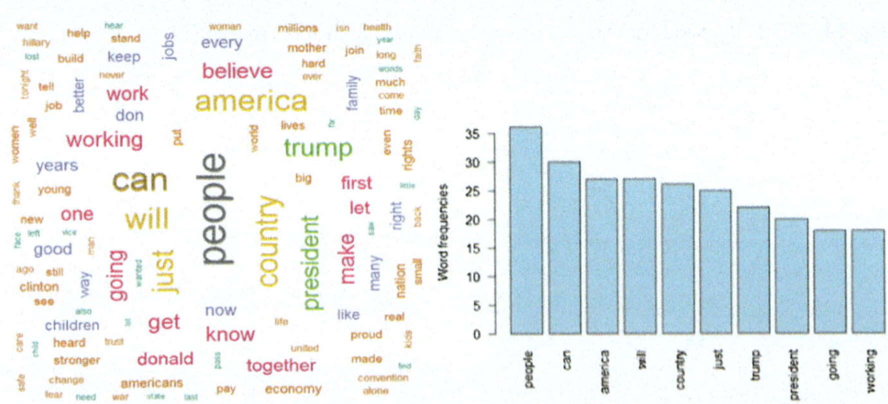

Fig. 17 Word cloud and bar chart for Clinton's word usage in Democratic Convention speech

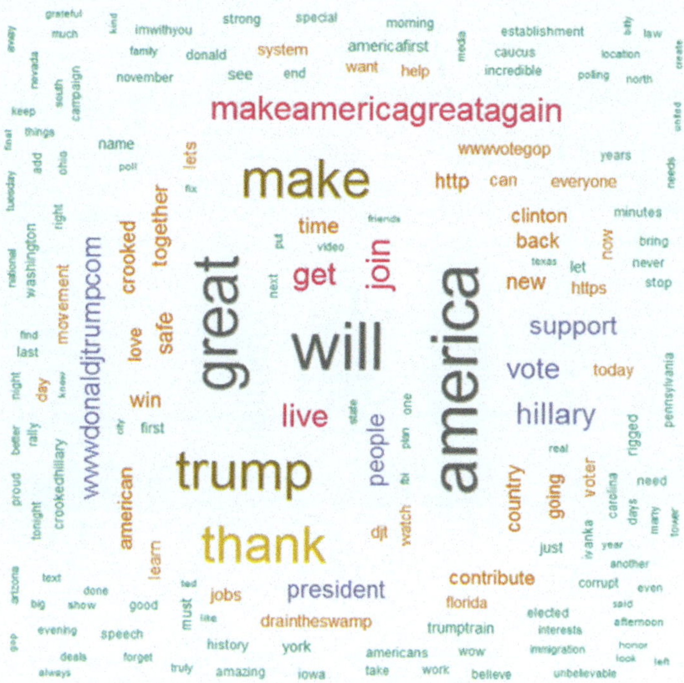

Fig. 18 Word cloud for Trump's word usage in his Facebook posts

Analysis—While Hillary too focused on criticizing Trump and his views, the major missing that comes out is that she does not seem to talk on the most pressing issues like that of job and immigration much which Trump majorly stressed on. This could be one of the reasons for her failure.

Trumps' Facebook posts wordcloud (Fig. 18)
Clinton's Facebook posts wordcloud (Fig. 19)

5 Conclusion and Insights Obtained

From the study conducted, some astonishing insights were revealed which were also true to the present-day facts. Hence, the social parameters that we considered for political event detection were totally justified to visually analyze the temporal change in candidate popularity. Even though in 300+ election polls and general opinion favored Hillary Clinton as the winning candidate, it was Donald Trump who succeeded in turning a large segment of non-aligned voters to his side through Facebook micro-advertising and wins elections. These facts were revealed with the data visualization approach we used to graphically compare the candidate's election

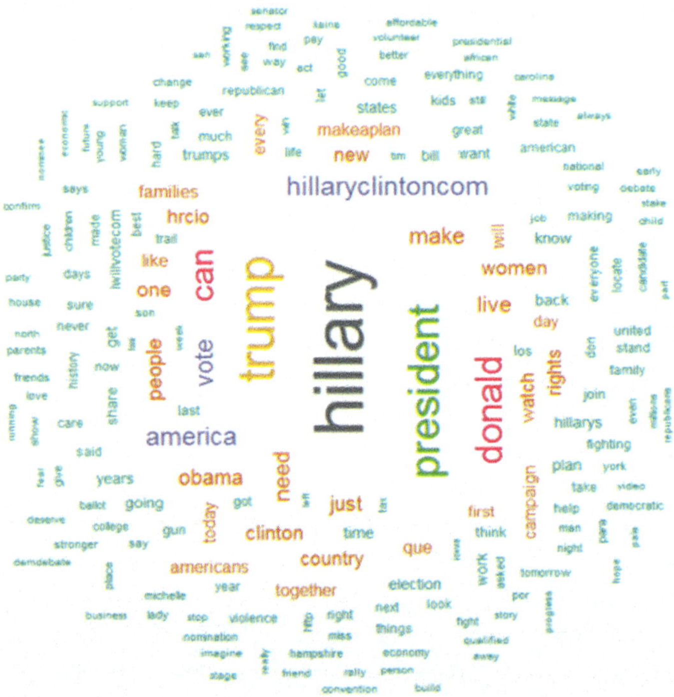

Fig. 19 Word cloud for Clinton's word usage in her Facebook posts

statistics. Facebook ads helped Trump win the poll and shaping public opinion which is clearly visible through insightful graphs generated as part of the study. All this beautiful analysis was made possible through data visualization techniques which again emphasizes its importance in our data-driven society. The elections were just one use case, and the data visualization techniques have enormous potential and scope. Data visualization provides an overall sense of the data such that the data speaks for itself. The same visceral response is rare when we just bluntly read the same data.

Evaluation of IoT Data Visualization Tools and Techniques

Suresh K. Peddoju and Himanshu Upadhyay

Abstract Internet of things (IoT) is a fully proven technology in the era of automation. IoT is a connected network of embedded systems with sensors and actuators. IoT generates huge volumes of data due to large number of implanted IoT devices everywhere. This generated data needs to be processed and analyzed to optimize operations and facilitate decision making. Data analytics plays a vital role in decision making. IoT has its applications in several areas: environmental monitoring, infrastructure management, manufacturing, energy management, medical and healthcare systems, building and home automation, transportation and many more. In every IoT application, a large amount of data has been generated with variations in it. Analysis, optimization and visualization of such huge data require smart tools and technologies. For example, some data requires specific algorithms to build models as a classification, whereas others require clustering and anomaly detection. The data visualization tools and techniques available for IoT data are very useful to get a better understanding of IoT, its framework, functions, and missions. There is still need for research and literature about data visualization tools and techniques for IoT and the challenges related to it. In this chapter, we have included various open-source commercial tools and techniques available in the market. We have also studied the benefits and challenges of existing tools. We analyzed and evaluated the suitability of existing tools and capabilities to gain leverage and support for IoT data visualization.

Keywords Internet of things · Machine learning · Data analytics · Data visualization · Tools and techniques

---

S. K. Peddoju (✉) · H. Upadhyay
Applied Research Center, Florida International University, Miami, USA
e-mail: speddoju@fiu.edu

H. Upadhyay
e-mail: upadhyay@fiu.edu

S. M. Anouncia et al. (eds.), *Data Visualization*,
https://doi.org/10.1007/978-981-15-2282-6_7

1 Introduction

Most large- and small-scale industries are paying attention to the Internet of things (IoT) due to the significant increase of smart devices in everyday life [1]. This has led to widespread use of applications in different domains; therefore, there is an increase in the potential growth of IoT devices in these areas. For the last five years, the usage of IoT in different domains led to a drastic increase in the potential economic rate. According to McKinsey, by 2025 there will be an annual increase of $2.7 trillion to $6.2 trillion [2]. Such an increase transforms these enterprises into a digital business, and a need to design new models which satisfies customers and maintains a good rapport between the customers and employers is created. However, implementation of IoT applications specifically to these industries to obtain a right solution is a challenging task. Gartner reports that by 2020, approximately 65% business industries will adopt IoT products into their business models [3]. IoT produces huge amounts of data that should be analyzed, made understand, and visualized for these industries.

The majority of industries produce vast amounts of data, and as a result, there will be a dramatic increase in data on the Web. It is a tough task for the employers to analyze, explore, and visualize the bulk amounts of data. For this, there is a need for visualizing enormous data in a better way to access and understand easily; usually, data is represented graphically to provide effective visualizations. This process of making good decision making with graphs to better understand data can be called data visualization, information visualization, or scientific visualization [4]. This helps to discover patterns, comprehend information, and make informed opinions. Customers and employers are guided with visualizations to inform their physiological movements using this data visualization.

It is better to visualize the data in a graphical manner, as it is an effective tool that helps people understand large amounts of data in a short amount of time, which accelerates their daily business. Business organizations usually demand more visualization to better understand their end users in order to expand their business as well. They need this data to provide insights that can better promote their products and business to these end users. Based on the visualizations made with the data, they can explore more efficient solutions to expand their business.

There are a variety of ways to change data visualization based on the interactions users have made with the data. Suitable analytics are needed due to this, and it provides better insight into the organization. There is an acute need of tools to enhance the ability of analyzation of data for these organizations to make quick and informed decisions. These decisions are based on the visuals given by the data visualization as either graphs, charts, or reports [5]. Again, this data visualization is represented in pictorial form to make information easier and quicker to understand and can help provide solutions that need to be implemented. The advent of computer graphics has helped shaped this data visualization by presenting large and complex data in an innovative way.

This chapter has provided a brief introduction to IoT, data visualization, the role of data visualization in IoT and shows some of the benefits of using tools and techniques in data visualization for IoT. Section 2 discusses the role of IoT in the present-day market. Section 3 discusses the importance of data visualization in different aspects and how it has been utilized effectively in different industries. Section 4 of the chapter covers the relation between IoT and data visualization within the context of data analytics in IoT and data visualization. Section 5 introduces tools and techniques used for general data visualization and further develops the significance of IoT-related visualizations. Section 6 concludes this chapter.

2 Internet of Things

IoT is a network of things made up of physical objects with embedded technology for proper communication, and sensing data from external entities, to have better interaction [6]. The connection of different things such as assets, processes, devices, and other personnel involved would capture data and its events to learn behavior and usage of IoT. Based on this, preventive actions are taken to convert commercial processes. IoT has capability of creating digital assets and businesses.

Greg Aimi, Gartner Research Director, says "Some of the earliest applications of autonomous technology for self-driving trucks are likely to be on the long hauls of highways." Michael Ramsey, another Gartner Research Director, says "Until clear leaders and standards begin to emerge, we will continue to see different alliances forming around the autonomous vehicle initiatives" [7]. This is an era of self-driving cars, which will use IoT technology. For example, a self-driving truck travelled all the way from 120 miles down the highways to reach Colorado with beer cans. It is an autonomous vehicle used for transportation of goods from one place to another under the supervision of driver from sleeper berth [5]. In the upcoming future, we can expect more autonomous vehicles for different means of transportation because self-driving cars can predict the route more accurately than a human can. Rather than navigating from general urban roadways, which will give a variety of difficulties such as street rules, route variations and many more, self-driving cars can pick the best possible route for the destination. "It's still the early days for driverless trucks, and Gartner predicts that by 2021, less than 1% of long-haul, over-the-road freight will be carried by autonomous trucks," says Aimi [7]. There are many applications that already exist in the market to exploit IoT-based driverless vehicles, but not commercially available. However, US officials are already saying that they are investing 4 billion dollars to upgrade new technology to bring the driverless vehicles on the roads for commercial purpose. But Michael Ramsey, Research Director, says "The biggest hurdle to autonomous vehicles becoming mainstream will likely be regulatory. Governments will need to feel comfortable with the rules put in place before these cars and trucks are released to the general public." At present, the development of autonomous vehicles differs with the technology developments, manufacturers, and chip companies. "The critical capabilities for automated driving cluster around sensing technologies, 3D mapping and data analytics, and algorithms for computer vision, localization and path planning," says Ramsey. "Until clear leaders and standards begin to emerge in each of these areas, we will continue to see different alliances forming around the autonomous vehicle initiatives of leading car companies." [7].

One of the applications of IoT is the connected home or smart home. A smart home is a connected network of things which can monitor the house; every device in the house is monitored, and even a gardening water system can be identified within the smart apps. We can also forecast the weather, see the intended water supply requirements, and much more can be done by applying analytics on the existing data. "The technologies and commercial effects of the connected home could have a wide-reaching impact on the role of CIOs [Chief Information Officer], depending on the company and products," says Nick Jones, Vice President and distinguished analyst. "However, the connected home also represents an opportunity for CIOs to be involved in the implementation and operation of new products and strategies. CIOs could be directly or indirectly impacted by the smart home, depending on whether the company is creating the connected products and services or determining how other companies' products will affect security." [7]. But they need to invest much more several hundred dollars to get a truly connected home, including a virtual personal assistant (VPA)-enabled wireless speaker, a smart lighting kit, door and window sensors, smart locks, and home monitoring cameras.

We can see many more applications coming up with IoT-based designs where things become customers. They can do all the activities that humans can and also conduct business far more efficiently. IoT is reserving spots in the nearest charging station for electric vehicles and also paying for these reservations based on user requirements. They have the intelligence to schedule a car maintenance. Recent updates in IoT have mad lots of differences in our lifestyle, such as an automated microwave that can set the temperature and time needed to reheat food based on what is inside it. Likewise, there are many more IoT applications which are grabbing the attention of the whole market. IoT has changed the provider–consumer relationship by replacing the existing workflow. These IoT devices are able to make a decision by themselves without the intervention of consumer and provider. Don Scheibenreif, Vice President and Distinguished Analyst [8], says "This may seem far-fetched, but it's not, and organizations that fail to build for such eventualities are at risk of obsolescence."

Innovation has made a pathway to identify Internet-connected things and also to identify themselves. Things can communicate with other technologies, customers, and businesses. IoT has made human lives easier by taking care of every task without much human interaction needed. By 2020, IoT devices are outnumber humans by four times and will create new dynamics in all services.

3 Data Visualization

Data can be represented in different formats such as pictorial, graphical, and other. This representation is called data visualization. It enables decision makers to see data analytics presented visually, so they can grasp difficult concepts or identify new patterns in an efficient manner [9, 10]. With interactive visualizations, interactively changing what data you see and how it is being processed. It empowers the user to see and examine displayed information superficially, so they can get a handle on what the data is representing or recognize new patterns within that data. With intelligent observation, you can take the idea a stride further by utilizing innovation to go into outlines and charts for more detail, intuitively changing what information you see and how it is handled.

Making use of visuals to comprehend information has been there since long time, and during the seventeenth-century time period, people were using maps, diagrams, and pie graphs. After few years later, there were major changes in factual designs happened at the time of Napoleon's attack of Russia [11] identified by Charles Minard. The guide delineated the extent of military power that Napoleon had and showcased their retreat from Moscow, attaching the data to temperature and time scales for a more inside comprehension of the occasion. It was an innovation that really lit the flame for data visualization, and with modern PCs, it is made more convenient to process a large set of information with pretty much good speeds. Nowadays, information and the understanding of information have revolved into a combination of science and workmanship that will change the culture of corporate style throughout the following couple of years.

Though information in content structure can be truly confusing, information shown in a visual configuration enables individuals to get significance from that information rapidly and effectively. You can uncover illustrations, designs, and connections that may go undetected if data was only given through speech. Data visualization can also be static or intelligent [12]. People have been utilizing static data visualization like outlines and maps for centuries. Intelligent data visualization, however, is a newer concept. It provides individuals a chance to go into the littlest of details in outlines and diagrams utilizing their PCs and cell phones and afterward intuitively change which information they see and how it is prepared.

Data visualization can also be used to identify certain areas that need improvement, give an in-depth report to customer behavior, attribute these behaviors to certain influences, make products better understood, better advertise to consumers, and predict market trends based on already existing data.

Most analytics plans are used to help basic leadership and fill in as apparatuses that increase comprehension. In planning and building data visualization model, one must be guided by how the user will perceive these connections. Data visualization is something beyond speaking only to numbers; it includes choosing and reconsidering the numbers on which the analyzation is based [13]. The same authors of this chapter present different methods to utilize the data of sensors, diabetes, and childhood pneumonia to represent the data analytics of predictions from the data [14–17]. Representation of data is a critical part of software engineering and has a wide scope of utilizations [18]. A few applications have explicit instruments that break down individual datasets in numerous fields of medicine and science [19]. Public health is the capacity to investigate and exhibit information in a reasonable way, which is the basic to achieve general well-being. Health analysts need valuable and innovative apparatuses to help their work [20]. Security is imperative in cloud-based therapeutic data analytics. Open any medical or health magazine today, and you will see a wide range of graphical portrayals. Renewal energy is the calculation of vitality utilization contrasted with the generation, which is critical for ideal arrangement [21]. Environmental science is ecological administrators that are required to settle on choices dependent on exceedingly complex information, which requires perception. Representation applications inside connected natural research are starting to rise [22]. It is attractive to have one distinct project for showing results. Fraud detection is the data representation that is vital in these times of extortion. A fraud agent may utilize data analytics as a proactive discovery approach, utilizing it to see patterns that may propose deceitful actions [23]. Library decision making allows data visualization software to permit administrators the adaptability to all the more likely oversee and present data gathered from various sources. It gives them the ability to show data in an inventive, convincing manner [24]. Visualization of library information features obtaining choices, future library needs, and objectives. Experts of data visualizations can help understudies, personnel and scientists imagine their information [25]. A few data visualization calculations and related programming have been created already, and these products empower clients to translate information more quickly than any other time. ManyEyes from IBM, SmartMoney for securities exchange, Insights from Facebook Corporation, Visual Analytics from SAS, and Thoth from California Institute of Technology, Tableau, and TOPCAT are a few of these products [26, 27]. They make data visualization simple to translate and fast to create. Each instrument has its own great highlights and confinements. Perception of an expansive scale multidimensional informational collection can be joined with new methodologies of interfacing with PCs utilizing the Web application as an administrator.

Extensive, time-changing datasets present extraordinary tests for data visualization due to the huge information volume. Continuous data visualization can empower clients to proactively react to issues that emerge. The movement age approach is utilized for intuitive investigation procedure of time-fluctuating information. It envisions fleeting occasions by mirroring the organization of narrating strategies [28]. Clients vary in their capacity to utilize data visualization and settle on choices under hard-time limitations. It is difficult to evaluate the value of the data visualization method. This is the purpose behind having a large number of visualization calculations and related programming [29]. A large portion of these products have not exploited the multi-contact associations and direct control capacities of newer gadgets. Large set of information, structured and unstructured, presents a one of a kind arrangement with difficulties for creating visualizations. This is because of the speed, size, and assorted variety of information. Other issues identified are involved with execution, operability, and degree of discrimination which challenge substantial data visualization and investigation [30]. It is troublesome and tedious to make an extensive informational index. It is additionally hard to choose what visual may be the best to utilize.

Data visualization is the way toward speaking to information in a graphical or pictorial route in an unmistakable and compelling way. It has risen as an incredible and generally relevant device for breaking down and deciphering expansive and complex information. It has turned complex data into speedy, simple methods for passing on ideas in a general manner that is easy to understand. It must discuss complex thoughts with clearness, exactness, and productivity. These advantages have enabled data visualization to be helpful in numerous fields of study.

4 Internet of Things with Data Visualization

IoT and data remain naturally connected together [31]. Data is devoured and created continuously, developing at a regularly growing rate. These developments in data accumulation reaching IoT appropriation as there will be a vast number of IoT gadgets nearly 30.73 billion by 2020. IoT is a network of few devices, systems, innovations, and other things to achieve a shared objective. There are assortments of applications made up of IoT being utilized in various ways and have prevailed with regard to giving gigantic advantages to their clients. The data generated from IoT gadgets turns out to be significantly worth it only if it gets exposed to analysis, which brings data analytics into the picture. Data analytics is characterized as a procedure, which is utilized to look at of all shapes and sizes of data sets with changing data properties to separate important conclusions and significant insights [32]. These ends are for the most part represented as patterns, examples, and insights that guide business associations in proactively interacting with information to actualize compelling basic leadership decisions.

Data analytics has a basic undertaking to do in the advancement and accomplishment of IoT applications and ventures. Data analytics tools will empower forte units to make effective usage of their datasets as explained in the focuses recorded underneath.

- **Volume**: There are immense bunches of datasets that IoT applications make utilization of. Business associations need to deal with these expansive volumes of data and need to break down important insights. So called datasets alongside continuous information can be broken down effectively and productively with data analytics algorithms.
- **Structure**: IoT applications include datasets that may have a fluctuated structure: unstructured, semi-structured, and structured datasets. There may likewise be a huge distinction in the data formats and types. Analytics will enable to break down these differing datasets utilizing automated tools and techniques.
- **Driving Revenue**: The use of data analytics in IoT hypotheses will empower the claim to specialty units to get a comprehension into client inclinations and choices. This would incite the improvement of organizations and offers as per the client's solicitations and desires. This will improve the incomes and advantages earned by the organizations.
- **Competitive Edge**: IoT is a stylish articulation in the present time of advancement, and there are different IoT application architects and providers present in the market. The use of data analytics in IoT markets will give a forte unit to offer better administrations and will have an aggressive edge in the market.

There are distinctive sorts of data analytics that can be used and applied in the IoT speculations to gain advantages [33]. A portion of these sorts have been listed and portrayed underneath.

- **Time Series Analytics**: As the name proposes, this kind of data analytics relies upon time-sensitive data which is separated to reveal related patterns. For instance, IoT applications, atmosphere evaluating applications, and well-being detecting frameworks can benefit from this kind of data examination strategy.
- **Streaming Analytics**: This type of analytics can also be called stream handling, and it dissects enormous in-movement datasets. Constant data streams are broken down in this procedure to distinguish critical circumstances and prompt activities. IoT applications dependent on financial transactions, air fleet tracking, traffic analysis, and so on can profit by this strategy.
- **Prescriptive Analysis**: This sort of analytics is the mix of explaining and perceptive investigation. It is associated with fathom the best walks of move that can be made in a particular condition. Business IoT applications can make usage of this kind of data analytics to increase income. There have been circumstances, wherein IoT markets have gigantically profited by the application and the usage of data analytics. With the change and movement in development, there are rising zones in which data analytics can be associated in relationship with IoT. For instance, advertising can be finished by applying data analytics to the product. IoT research will similarly allow the extended security and reconnaissance capacities through video sensors and usage of data analytics strategies.
- **Spatial Analytics**: This is the data analytic method that is used to dismember geographic patterns to choose the spatial association between the physical items. For instance, territory-based IoT applications and smart halting applications can benefit from this kind of analytics.
- **Healthcare Perspective**: Healthcare is one of the prime areas of each nation and the usage of data analytics in IoT-based healthcare applications can provide a leap forward. Doing so, it may decrease healthcare costs, improve telehealth monitoring, and remote health services, increased diagnosis and treatment can be accomplished using the same.

The utilization of data analytics will be progressed in the area of IoT to increment improved incomes, forceful expansion, and customer commitment. By cooperating with the right framework accessory, associations can couple data analytics with IoT to go through information for picking a high ground.

IoT creates a lots of data is also called Big Data which is useless in itself, except if it is changed over into an arrangement that is understand, analysis which is presented in [34]. Only interfacing IoT gadgets and gathering information is never again a big accomplishment. Rather, what makes a difference is the way you use the IoT data in your business. It is a guarantee that visualization is a key piece of the IoT data utilization.

Individuals have the misconception that visualization is only an accumulation of diagrams and infographics. However, visualization requires context framed by the analysis and comprehension of data, and above all else the meaning of necessities and applications. It is likewise imperative to change IoT data into a helpful arrangement to pick up data that will profit a business.

The data from a sensor can be very effectively and naturally imagined into a time series utilizing a traditional line chart, for instance. There is a place and requirement for such visualization of the data produced by an individual sensor. The most widely recognized motivation behind why the data produced by a solitary sensor should be imagined is presumably to screen the status, activity or area of a gadget, or toward the start of a venture just to make the data noticeable.

IoT data forces a few prerequisites on the innovation of the visualization tool utilized. On account of monitoring type visualization, real-time information is an outright necessity. Another reasonable necessity includes the volume of data: The innovation must empower the visualization, accumulation, and filtering of a huge volume of data points.

In an ideal circumstance, visualizations that are made for checking needs turned out to be less critical as the project continues and develops. Breaking point esteems and caution cutoff points can be indicated dependent on an analysis of the IoT data or simply dependent on the necessities, which implies that the center changes from dynamic checking to superficial observations where focused data can be sent dependent on the data at whatever point explicit measures are required. The info, cautions, and explicit tenets improve operational productivity and decrease the time utilized for manual monitoring.

The distinction among monitoring and alarms can be depicted with a straightforward model concerning the refrigeration device in supermarkets. The temperature in the refrigerator should remain between explicit farthest point esteems. Instead of physically confirming the temperatures by physically setting to each fridge step by step, monitoring can be utilized to check the temperatures from one screen at indicated esteems. The perfect case is that a gadget will issue an alert if the worth remains underneath or outperforms beyond as far as possible value. The alert is showed up in the terminal gadget of a person who will make the vital activity in order to keep the loss of products, to check the condition of the gadget, and to have it fixed, if essential. Sometimes, the individual simply needs to go to the refrigerator to close the doors that a customer has left open.

The extraordinary advantages offered by IoT data cannot be accomplished specifically by visualizing or monitoring the IoT data. Rather, the advantages can be accomplished by dissecting data from devices and sensors and utilizing the IoT data to determine indicators that allow for control of the tasks to improve the business, to lessen energy consumption, or to make the activities more environmentally friendly, for instance.

The advantages that can be accomplished by breaking down information from devices and sensors and uses the IoT data to infer business indicators are great. Staying with the case of refrigerators in a supermarket, you can easily calculate the difference of temperature inside a refrigerator. The lower the change, the less energy the device should consume. You can set a target level for the change and concentrate the difference between the values of refrigerators, for example.

More often, definitive advantages can be accomplished by combining IoT data and the indicator got from it with the organization's other information. Let us consider a supermarket: You introduce sensors on every aisle and at the entrance that give information on what number of individuals visit every aisle. The quantity of visitors is as of now an indicator which portrays what number of visitors there are at a particular time of day, why individuals go to the store, what is the most visited aisle, what number of individuals in all visit the cosmetics products aisle, and so forth. But imagine a scenario in which you combine the information on the quantity of customers with deals information. You could without much of a stretch ascertain what number of the customers who visited a particular aisle really purchased something and how much cash they spent. Is there an aisle that individuals visit without purchasing anything? By consolidating data, you can acquire totally different data about the business, which enables you to improve your tasks. Simply associating IoT gadgets and gathering information are not a big accomplishment. Rather, what is important is the means by which you use the IoT information in your business. You can utilize IoT information, especially on the off chance that you consolidate it with already existing information, to find new business examples and patterns. You can utilize the data to build up your activities, advance expenses or even make new administrations and income.

5 Data Visualization Tools and Techniques for IoT

The most ideal approach to exhibit the data is visually. There is a term for that: data visualization. Data visualization considers any type of graphic content visually communicates data to the viewer. There is a saying; a single picture is worth a 1000 word. This section describes data visualization tools and techniques required for IoT and the purposes of these tools [35].

5.1 Different Charts for Data Visualization

One of the struggles that hinders reporting and analysis is understanding what sorts of diagrams to utilize and why. That is in light of the fact that picking the wrong visual aid or simply defaulting to the most outstanding kind of data visualization could make confusion with the viewer or lead to mistaken data interpretation.

To make graphs that clarify and give the correct canvas to analyze, you should initially analysis the reasons for drawing chart. The following are different charts and graphs shown in Fig. 1 that are usually used in the present-day market.

Fig. 1 Different graphs, charts, and maps for data visualization

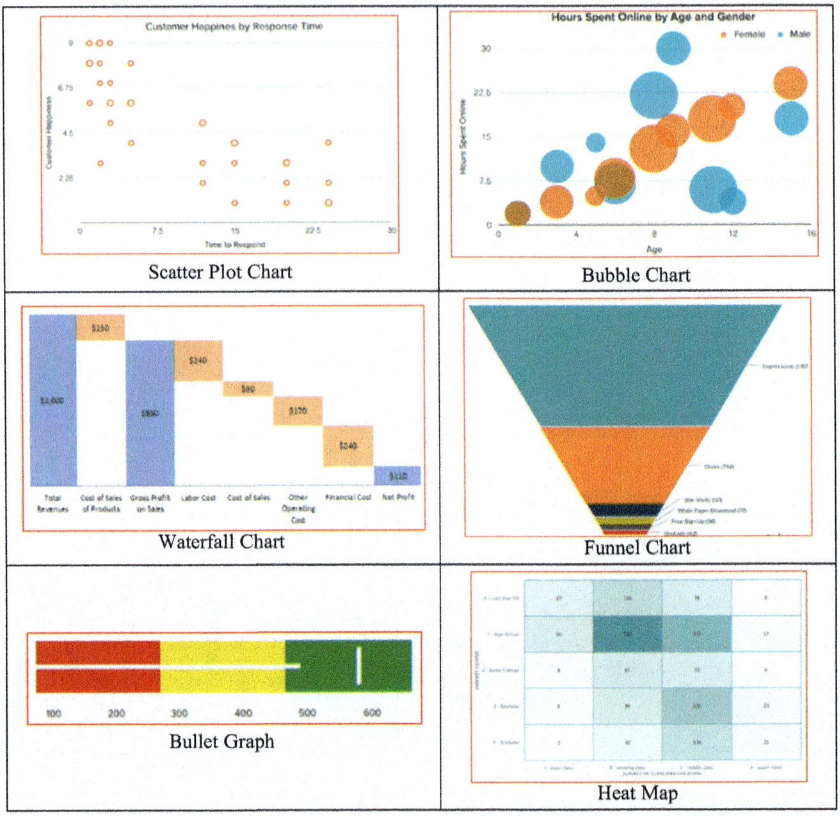

Fig. 1 (continued)

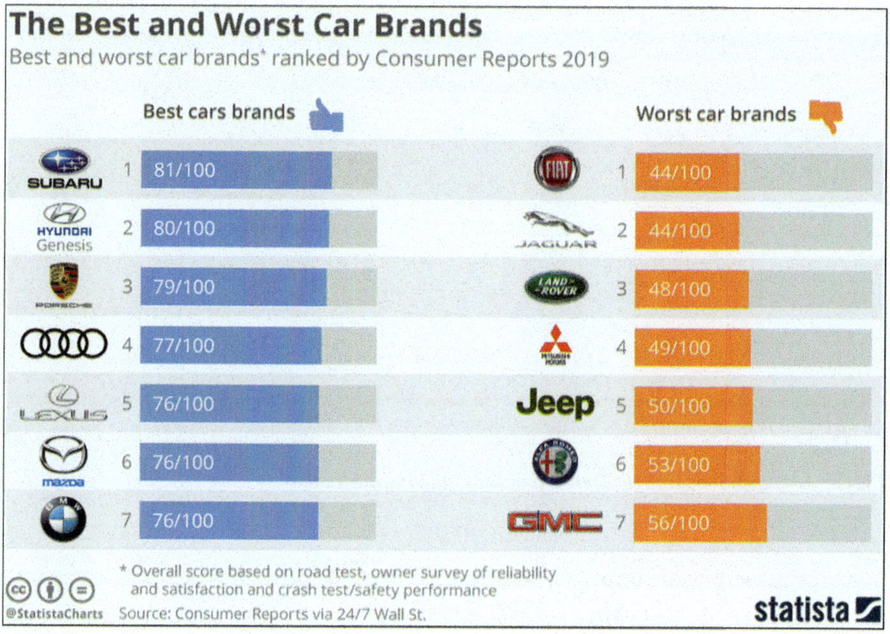

Fig. 2 A snapshot of *Statista tool for data visualization*

5.2 Tools for Source Credible Data

The following mentioned tools are the most popular source for credible data and will make dazzling visualizations.

Statista is a statistics, studies, and forecasts portal concentrated on statistical surveying and estimation polling. Implied for industries and academics, this makes it simple to discover solid market data dependent on industry. A snapshot is shown in Fig. 2.

Google Trends gives you information on what individuals are scanning for, how trends change after some time, and how search intrigue varies by region, district, nation, etc. It is anything but difficult to scan for explicit trends or essentially browse current trending topics. A snapshot is shown in Fig. 3.

Zanran is a search engine structured explicitly for discovering tables, outlines, and charts online. It works by first looking at pictures found on the Web, not content. A snapshot is shown in Fig. 4.

Pew Research Center, one of the main research organizations in the USA, distributes huge amounts of data and information on public opinion, social issues, and socioeconomics in the USA also, around the world. A snapshot is shown in Fig. 5.

Social mention is similar to Google Trends is a search and analysis tool that enables you to screen data generated by user based on patterns on the Web. It can

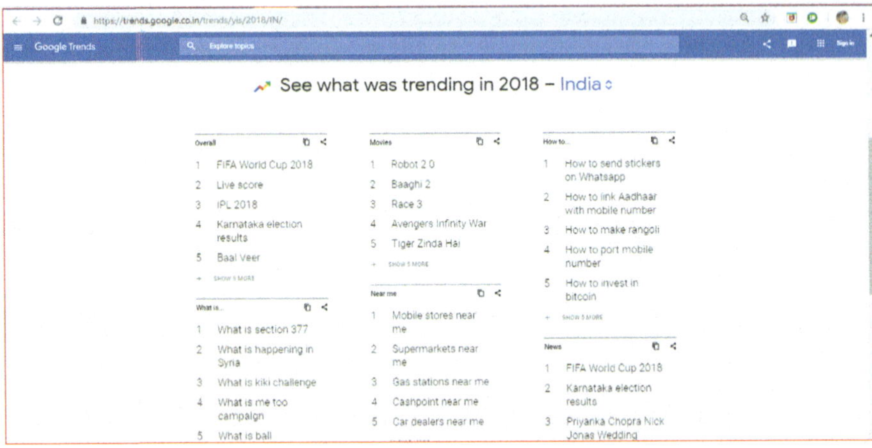

Fig. 3 A snapshot of *Google Trends tool for data visualization*

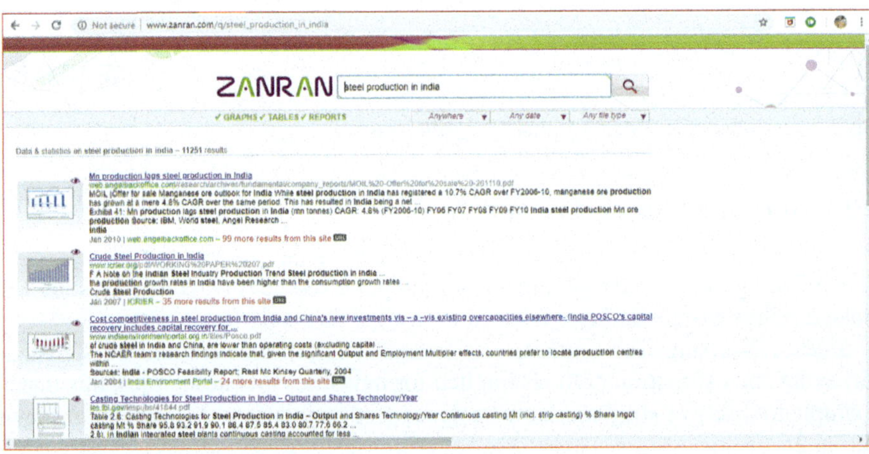

Fig. 4 A snapshot of Zanran *tool for data visualization*

be used to monitor the brands of individual or organizations. It is a powerful tool to analyze the features and trends. A preview appeared in Fig. 6.

ThinkwithGoogle is a Google's tool to find the latest data trends for marketers, and this tool is specially designed for marketers for keeping them up to date. A snapshot is shown in Fig. 7.

HubSpot Research is the place where new and original reports, statistics, charts, and thoughts were published. A snapshot is shown in Fig. 8.

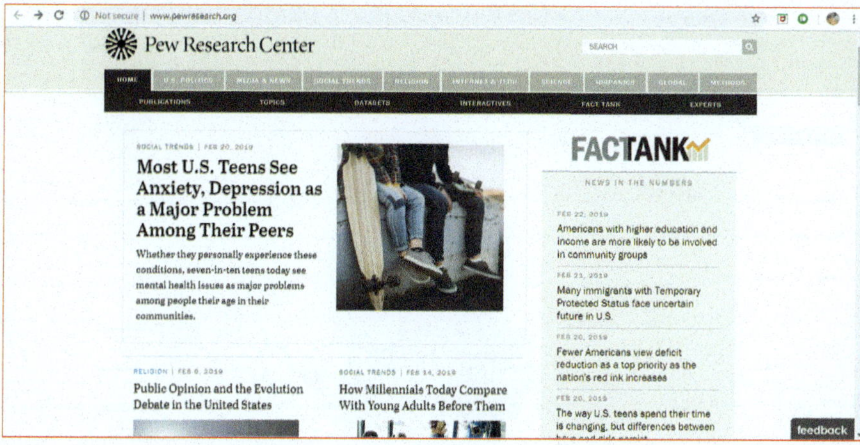

Fig. 5 A snapshot of Pew Research Center *tool for data visualization*

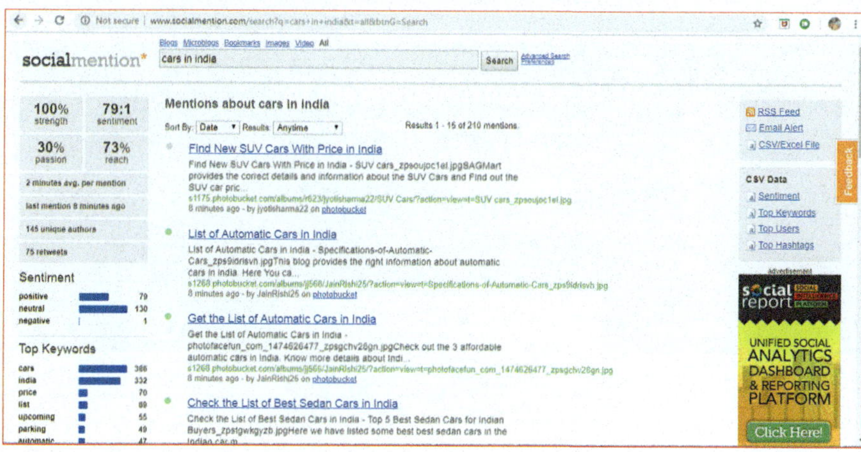

Fig. 6 A snapshot of social mention *tool for data visualization*

5.3 *Tools for Creating Data Visualizations*

Since you presently realize where to find reliable information, it is an incredible chance to start thinking about how you are demonstrating that information that functions for your targeted audience. At its core, data visualization is the route toward changing basic statistical data points into an absorbable picture—it may be diagram, graph, timeline, map, infographic, or other sorts of visual. While understanding the theory behind data representation is a basic, you moreover need the tools and

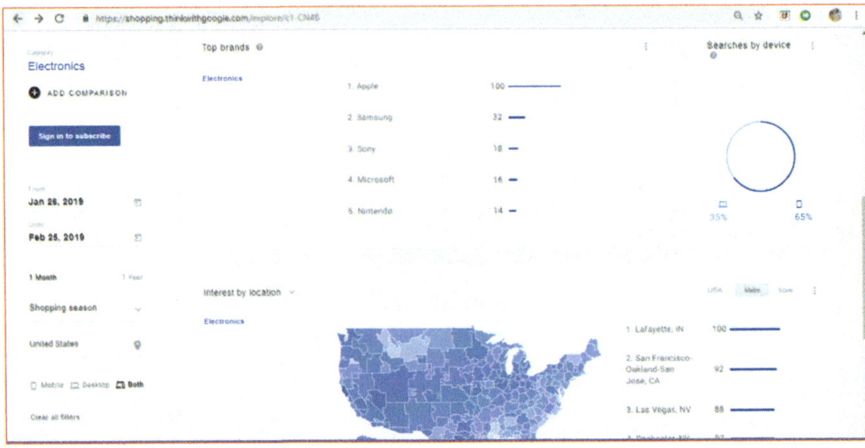

Fig. 7 A snapshot of *Thinkwithgoogle tool for data visualization*

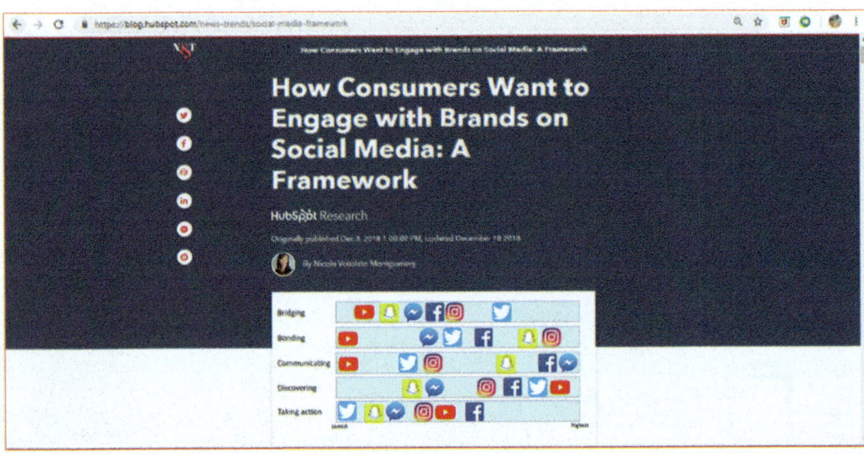

Fig. 8 A snapshot of HotSpot *tool for data visualization*

resources to make propelled data visualization possible. Underneath, we are mentioning powerful tools to scrutinize, bookmark, or download to make designing data visuals which are more valuable for any business or organization.

Infographics are an incredible method to translate your data by transforming it into something that tells a visual, noteworthy story. In the event that you have practically zero plan understanding, ***Infogram*** is an incredible tool for you. It offers distinctive infographic formats and tools for modifying your infographic. You can utilize charts, diagrams, maps, pictures, and symbols to truly zest up your data and make it visually appealing. A snapshot is shown in Fig. 9.

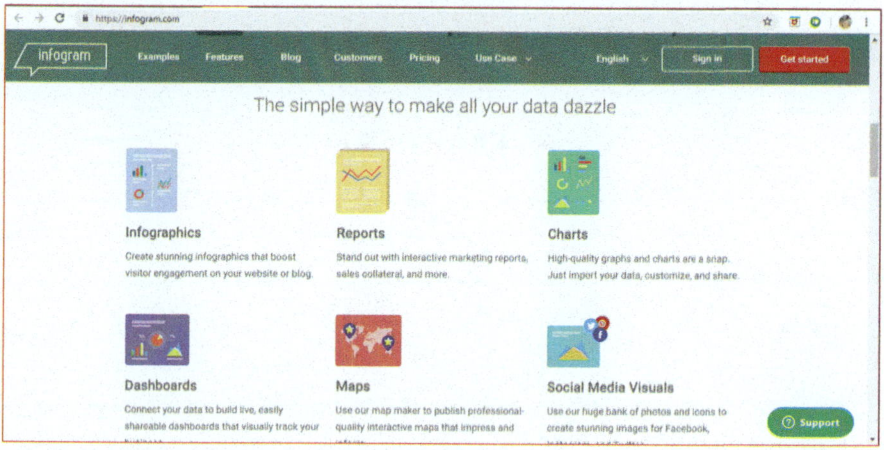

Fig. 9 A snapshot of *Infogram tool for data visualization*

A really sophisticated data visualization tool being used by most of the market is ***Tableau***. This tool interacts with other data tools such as Excel. It will convert the bare information into wonderful data visuals and made it easy for the people. This tool is bit expensive but most powerful to handle volumes of data and had big set of data analytics. A snapshot is shown in Fig. 10.

Tableau being used by many organizations and their data visualizations is available in tableau public. A snapshot is shown in Fig. 11.

Trifacta integrates natively with Tableau Data Extracts (TDE), which allows huge set of analysts to make use of these two technologies. For example, PepsiCo Company uses these two technologies by making use of Trifacta for data handling and Tableau

Fig. 10 A snapshot of Tableau *tool for data visualization*

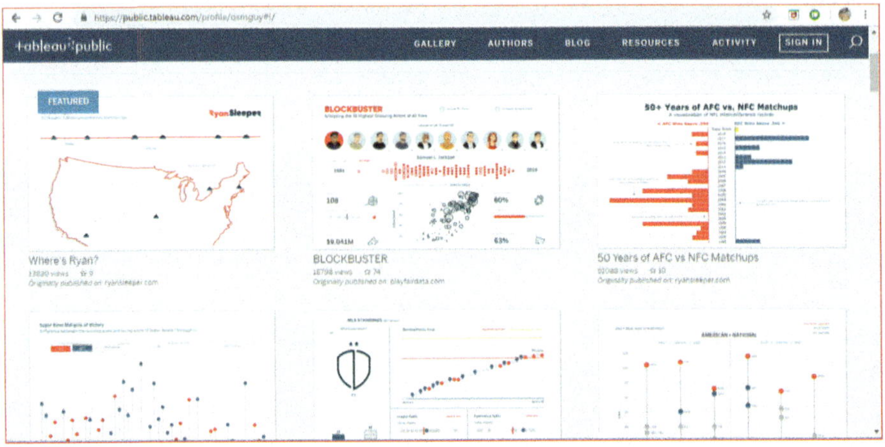

Fig. 11 A snapshot of Tableau public *tool for data visualization and representation*

for applying analytics on that data. With this step, the company has dramatically reduced time spent on data preparation and increased the overall quality of the data.

As shown in Fig. 12, PepsiCo builds reports and visualizations within no time and analyze the status. With this approach, they are reducing production time by 90%.

To visualize marketing performance data for presentations or reports to share with clients or peers, ***Databox*** provides standardized report templates for 50+ popular marketing software products including Google Analytics, HubSpot, AdWords, and Facebook ads. We can allow others to view and access your up-to-date data from many devices including a computer, via the Databox mobile app, a TV, or even an Apple Watch. A snapshot is shown in Fig. 13.

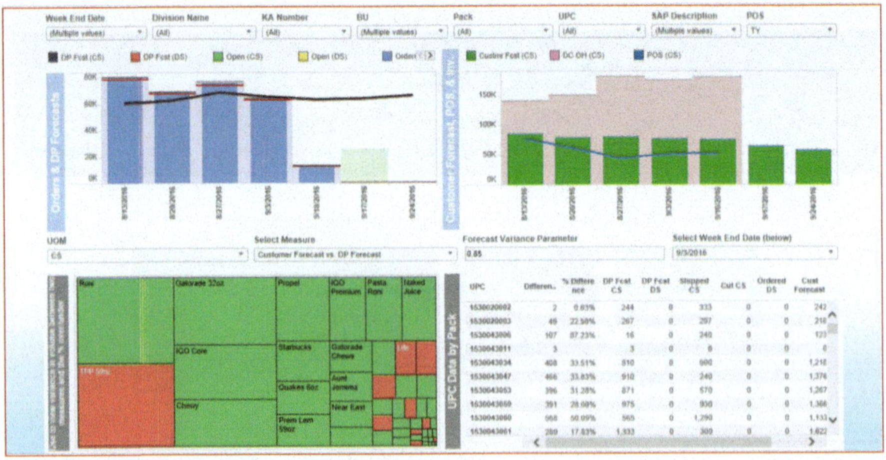

Fig. 12 A snapshot of Tableau *tool for PepsiCo Ltd Co., data visualization*

Fig. 13 A snapshot of Databox *tool for data visualization*

ZingChart is JavaScript-based library to design charts, graphs, maps, and other visuals to place it in different Web sites and blogs used by different users. This tool is easy to use and flexible to adopt. Its capability is to design world-class visuals compatible with any device and any screen. A snapshot is shown in Fig. 14.

Google provides an API to create custom visuals which can capable of embedding in any Web page. The name of the tool is ***Google Charts*** which is similar as Google Sheets. A snapshot is shown in Fig. 15.

Piktochart is a tool similar to Infogram which is used to create and customize infographics within its templates. By using this tool, designer can create custom and wonderful infographics. A snapshot is shown in Fig. 16.

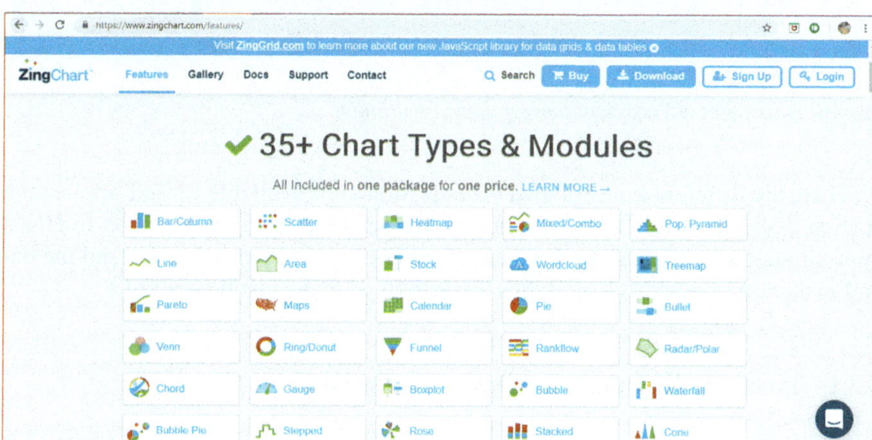

Fig. 14 A snapshot of ZingChart *tool for data visualization*

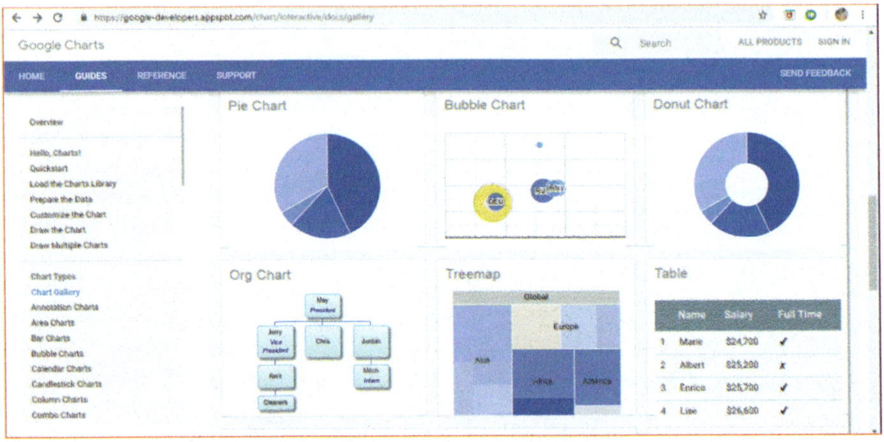

Fig. 15 A snapshot of Google Chart *tool for data visualization*

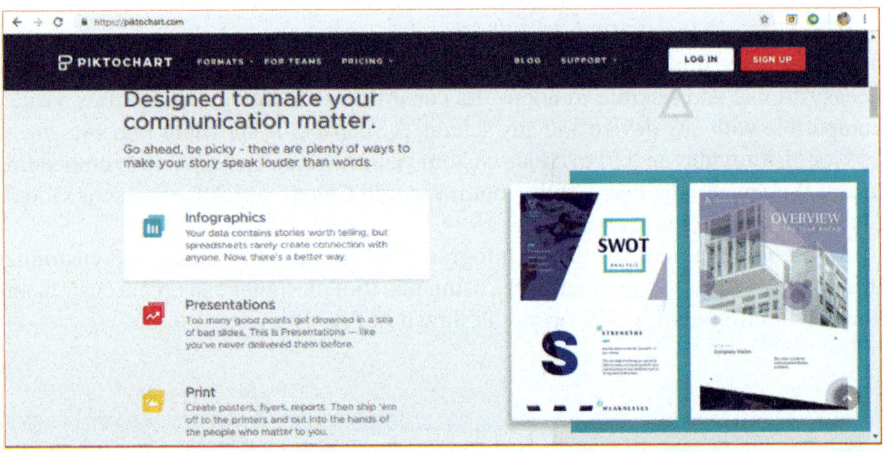

Fig. 16 A snapshot of Piktochart *tool for data visualization*

There are different kinds of data visualization tools available for IoT applications such as Kibana, Power BI, and Grafana. As per requirements, we need to choose appropriate tool and analyze the data. These tools also used to understand the data and make use of advanced analytics algorithms to make insights out of it.

5.3.1 Power BI for Real-Time Data Visualization

Power BI is the tool similar to Tableau for representing data related to IoT applications. Initially, Power BI was a comprehensive commercial knowledge base that

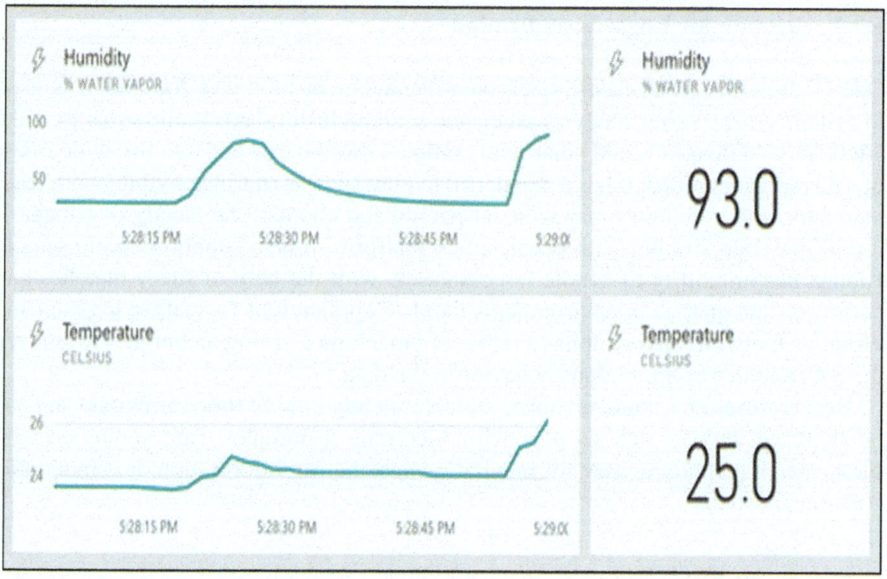

Fig. 17 Image Credit: Microsoft Power BI Blog

empowered organizations to picture a wide scope of datasets. It has a complete rundown of mixes, and it can pull information from numerous sources like Excel, Google Analytics, Salesforce, and Web-based life stages. Likewise with most Microsoft items, Power BI is very natural and has a sensible expectation to learn and adapt. That being stated, Power BI is a paid administration.

Benefits of Power BI will able to process different kinds of data such as streaming as well as static data, magnificent data visual types, different query optimization techniques and many more. A snapshot is shown in Fig. 17.

5.3.2 Grafana for Metrics Visualization

A powerful data visualization and analytics tool called ***Grafana*** is advanced tool to visualize and analyze time series data. Grafana is similar to Power BI which will work with different kinds of data and support wide range of data visualizations. It can be used to build dashboards, handle, and manage data. This tool is vastly used to monitor CPU including system health. This tool is widely used in IoT and related fields to analyze the time series data and visualize them.

Its rich set of features include building smart dashboards, messaging mechanisms, alerts and notifications, and handling streaming data. It uses different querying techniques to handle different data sources. With the usage of Grafana, it is easy to handle different data sources, metrics data, messaging mechanisms, filtering to custom visuals, personalization of dashboards and many more.

5.3.3 Kibana for Logs' Visualization

One of the tools from elastic stack data management toolkit is ***Kibana.*** It is used to visualize time series analysis on elastic search cluster data. It allows us to make effective visualizations and smart and complex dashboards and can use wide range of data representations. It has a wonderful feature such as sharing anything with team members which includes management people and clients. The beauty of Kibana is it will integrate straight with elastic stack platform. It also supports elastic search. Kibana has an ability to handle large datasets with the help of fuzzy matches and performs data queries in environments like IoT applications to analyze log data and visualize them effectively. It has a rich set of machine learning techniques to analyze the data and detect anomalous behavior in the data.

Best features of Kibana includes, suitable for any kind of time series data and can be analyzed with its rich set of machine learning techniques, can be queried with fuzzy match technique, easy for integration and sharing information. It is integrated with elastic stack.

5.4 Platforms, Tools, and Libraries for IoT Data Visualization

DGLux is a platform for application development and visualization. This can be used by individuals or organizations to handle real-time applications. It has many features that can be dragged and dropped to develop user-friendly and data-driven applications and dashboards. The beauty of this tool is no need to write any single line of code to build such applications and dashboards.

One of the open-source platforms to create custom dashboards for any applications especially IoT applications is ***Freeboard***. It has an interface with which we can build real-time, interactive applications with just dragging and dropping required features from interface. Effective visualizations can be designed on it within no time due to its rich set of tools.

Nokia designed a platform named ***Here*** to provide location-based services with GIS and mapping mechanisms. It provides custom mapping services and technologies to different sectors.

IBM's Bluemix is an IoT platform which will connect multiple devices and sensors. It has numerous sets of API to provide visualizations on different devices and display formats.

Luciad brought a powerful tool for geospatial solutions that power the world's mission critical operations. It has ability to provide advanced analytics with effective visualizations on real-time location intelligence. It is been used for smart cities to safeguard digital infrastructure.

Hans Scharler's ***ThingSpeak*** is a platform to handle sensor data. This platform is an open source and supported by MATLAB can be used to analyze and visualize sensor data.

Apart from above-mentioned data visualization tools and techniques, there are various tools available in present market. But because of space constraints, we are not including them in this chapter.

6 Conclusion

Data visualization is a passionately debated issue for IoT at the present time. Most of the organizations fuse data focused activities to increase their overall operations and design new strategies. This brings data visualization into picture. IoT technology is widely used by customers and service providers which requires good feel of data visualizations. In this chapter, we identified different tools to figure out how to get insights from existing information. It is vital first to recognize objectives and needs to identify suitable tool from diverse data visualization platforms. In this chapter, we have included different open-source tools and techniques available in the market. We have additionally contemplated advantages and difficulties of existing tools. We have made an investigation to assess the appropriateness of existing tools and techniques to use and support IoT data visualization. We have likewise proposed research bearings for IoT state-of-the-art tools and techniques to get better visualization.

References

1. Özen, F. (2018). Internet of Things and Data Visualization—Exastax, Exastax.com, viewed 14 February, 2018. https://www.exastax.com/big-data/internet-things-data-visualization/.
2. Mckinsey. Where machines could replace humans—and where they can't(yet). http://www.mckinsey.com/business-functions/business-technology/our-insights/where-machines-could-replace-humans-and-where-they-cant-yet. Accessed July 25, 2016.
3. Gartner. (2015). Gartner says 6.4 billion connected 'Things' will be in use in 2016 up 30 percent from 2015 (online). http://www.gartner.com/newsroom/id/3165317.
4. Chen, C., Härdle, W., & Unwin, A. (2008). *Handbook of data visualization* (pp. 1–954). Berlin: Springer.
5. Mohammadi, M., Al-Fuqaha, A., Sorour, S., & Guizani, M. (2017). *Deep learning for IoT big data and streaming analytics: A survey*. arXiv preprint arXiv:1712.04301.f.
6. Chung, C., Chen, C., Shih, W., Lin, T., Yeh, R., & Wang, I. (2017). Automated machine learning for Internet of Things. In *2017 IEEE International Conference on Consumer Electronics, Taiwan (ICCETW)*.
7. Hung, M. (2017). Leading the IoT: Gartner insights on how to lead in a connected world. ebook Gartner digital.
8. Scheibenreif, D. (2016). Article: PDQ POS Top 10 Award Retail CIO Outlook.
9. Bikakis, N. (2018). Big data visualization tools. arXiv preprint arXiv:1801.08336.
10. Batista, A. F., Correa, P. L., & Palanisamy, G. (2016, October). Visual analytics improving data understandability in IoT projects: An overview of the US DOE ARM program data science tools. In *2016 IEEE 13th International Conference on Mobile Ad Hoc and Sensor Systems (MASS)* (pp. 349–354). IEEE.
11. Napoleon's Russian campaign: From the Niemen to Moscow, Napoleon.org. http://www.napoleon.org/en/Template/chronologie.asp?idpage=481959&onglet=1.

12. Chung, S., Suh, S., Park, C., Kang, K., Choo, J., & Kwon, B. C. (2016). ReVACNN: Real-time visual analytics for convolutional neural network.
13. Hohman, F., Kahng, M., Pienta, R., & Chau, D.H. (2018). Visual analytics in deep learning: An interrogative survey for the next frontiers. arXiv preprint arXiv:1801.06889.
14. Peddoju, S. K., Kavitha, K., & Sharma, S. C. (2016). Big data analytics for childhood pneumonia monitoring. Published in Edited Book "*Cloud computing systems and applications in healthcare*." USA: IGI Global Publisher.
15. Chaudhary, A., Peddoju, S. K., & Peddoju, S. K. (2016). Cloud based wireless infrastructure for health monitoring. Published in Edited Book "*Cloud Computing Systems and Applications in Healthcare*." USA: IGI Global Publisher.
16. Suresh Kumar, P., & Pranavi, S. (2017). Performance analysis of machine learning algorithms on diabetes dataset using big data analytics. In: *Proceedings of IEEE 2017 International Conference on Infocom Technologies and Unmanned Systems (ICTUS'2017)*, Dubai, United Arab Emirates (UAE) (pp. 580–585).
17. Suresh Kumar, P., & Umatejaswi, V. (2016). Diagnosing diabetes using data mining techniques. *International Journal of Scientific and Research Publications, 7*(6), 705–709.
18. Wang, H., Xu, Z., & Pedrycz, W. (2016). An overview on the roles of fuzzy set techniques in big data processing: Trends, challenges and opportunities. *Knowledge-Based Systems, 118,* 10–12.
19. Wolfe, J. (2015). Teaching students to focus on the data in data visualization. *Journal of Business and Technical Communication, 29*(3), 344–359.
20. Kilimba, T., Nimako, G., & Herbst, K. (2015, September). Data everywhere: an integrated longitudinal data visualization platform for health and demographic surveillance sites. In *Proceedings of the 6th ACM Conference on Bioinformatics, Computational Biology and Health Informatics* (pp. 551, 552).
21. Kumar, O., & Goyal, A. (2016). Visualization: A novel approach for big data analytics. In *Proceedings of the Second International Conference on Computational Intelligence & Communication Technology* (pp. 121–124).
22. Grainger, S., Mao, F., & Buytaert, W. (2016). Environmental data visualization for non-scientific contexts: Literature review and design framework. *Environmental Modelling and Software, 85,* 299–318.
23. Dilla, W. N., & Raschke, R. L. (2015). Data visualization for fraud detection: Practice implications and a call for future research. *International Journal of Accounting Information Systems, 16,* 1–22.
24. Murhy, S. A. (2013). Data visualization and rapid analytics: Applying tableau desktop to support library decision-making. *Journal of Web Librarianship, 7*(4), 465–476.
25. Brigham, T. J. (2016). Feast for the eyes: An introduction to data visualization. *Medical Reference Services Quarterly, 35*(2), 215–223.
26. Chen, C. (2010, July/August). Information visualization. *WIREs Computational Statistics, 2,* 387–403.
27. Laher, R. R. (2016). Thoth: Software for data visualization and statistics. *Astronomy and Computing, 17,* 177–185.
28. Yu, L., et al. (2010). Automatic animation for time-varying data visualization. *Pacific Graphics, 29*(7), 2271–2280.
29. Li, X., et al. (2015). Advanced aggregate computation for large data visualization. In: *Proceedings of IEEE Symposium on Large Data Analysis and Visualization* (pp. 137, 138).
30. Alton, L. (2016). 4 potential problems with data visualization. Datasciencecentral.com, viewed 20 March, 2018, https://www.datasciencecentral.com/profiles/blogs/4-potentialproblems-with-data-visualization.
31. Endert, A., Ribarsky, W., Turkay, C., Wong, B., Nabney, I., Blanco, I., et al. (2017). The state of the art in integrating machine learning into visual analytics. *Computer Graphics Forum, 36*(8), 458–486.
32. Keim, D., Andrienko, G., Fekete, J., Görg, C., Kohlhammer, J., & Melançon, G. (2008). Visual analytics: Definition, process, and challenges. In *Lecture Notes in Computer Science* (pp. 154–175).

33. Joseph, T. (2018). Role of data analytics in Internet of Things (IoT). Fingent article.
34. Sun, Y., Song, H., Jara, A. J., & Bie, R. (2016). Internet of things and big data analytics for smart and connected communities. *IEEE Access, 4,* 766–773.
35. Arockia Panimalar, S., Khule, K. M., Karthika, S., & Nirmala Kumari, T. (2017). Data visualization tools and techniques for datasets in big data. *International Research Journal of Engineering and Technology (IRJET), 4*(8), 1667–1672.

Data Visualization: Experiment to Impose DDoS Attack and Its Recovery on Software-Defined Networks

Bhargavi Goswami, Stanly Wilson, Saleh Asadollahi and Tony Manuel

Abstract The entire network is doing paradigm shift towards the software-defined networks by separating forwarding plane from control plane. This gives a clear call to researchers for joining the ocean of software-defined networks for doing research considering its security aspects. The biggest advantage of SDN is programmability of the forwarding plane. By making the switches programmable, it can take live instructions from controllers. The versions of OpenFlow protocol and the compatibility of programmable switches with OpenFlow were the stepping stone making software-defined networks thrashed towards reality. The control plane has come up with multiple options of controllers such as NOX [2], Ryu [3], Floodlight [4], OpenDayLight [6], ONOS [7] and the list is big. The major players are Java based which keeps the doors open for enhancement of features by the contributors. However, more is expected from the practicality of P4Lang programmed switches by bringing skilled people to the industry who can actually implement programmable switches with ease. The obvious reason for delayed progress in the area of software-defined networks is the lack of awareness towards data visualization options existing as of now. The purpose of writing this chapter is to throw light upon the existing options available for data visualization in the area of SDN especially addressing the security aspect by analyzing the experiment of distributed denial of service (DDoS) attack on SDN with clarity on its usage, features, applicability and scopes for its adaptabilities in the world of networks which is going towards SDN. This chapter is a call to network researchers to join the train of SDN and push forward the SDN technology by proved results of data visualization of network and security matrices. The sections and subsections show clearly the experimental steps to implement DDoS attack on SDN and further provide solution to overcome the attack.

B. Goswami (✉)
School of Electrical Engineering and Computer Science, QUT, Brisbane, Australia

S. Wilson
Department of IT, St. Vincent Pallotti College of Engineering and Technology, Nagpur, India

S. Asadollahi
ITTECS, Brisbane, Australia

T. Manuel
CHRIST, Bangalore, India

S. M. Anouncia et al. (eds.), *Data Visualization*,
https://doi.org/10.1007/978-981-15-2282-6_8

Keywords DDoS · SDN · Security · Denial of service · Floodlight · Mininet

1 Introduction

The network is expanding day by day and the current devices that are used for networking are becoming incapable of managing the flow of data from source to destination in an optimal and simplified way. The low-level language programmed devices which are closed to vendor-specific configuration not only cause limitations for innovation and implementations of new ideas for improved switching [8]. It also does not provide mechanism for automatically responding to wide ranges of events that may occur in the network. In order to respond to certain event in the network, the network operator must have to manually make the necessary changes in the network configuration by adding ad hoc scripts. The manual configuration of device frequently leads to misconfigurations of other event handling procedures.

Software-defined network (SDN) is a technology that facilitates to manage the network and permits to configure the network programmatically in order to improve performance and easy monitoring. SDN is meant to address the difficulties of the traditional static architecture of networks which is very much decentralized and complex [9]. The networks of today require more reliable and easy to fix even remotely with security [10]. In this context, the SDN comes into have better monitoring of the network and remote management. We admire the contributors for bringing the idea to reality where the major players are passionately pushing the research domain towards the progress overcoming the limitations and hurdles coming for making software-defined network. The work projected here is a part of the research project that tries to make a study on security aspects such as the distributed denial of service (DDOS) attack on the SDN controller and bring solution to the problem of DDoS attacks on SDN [11, 12]. Denial of service (DOS) is an attack launched on a network with a motive to make the system or the server down or function improperly. The purpose of DDoS is to make the system not available for the authorized or intended users and making them feel that system is busy. The attack can be launched temporarily or indefinitely. The SDN controller is flooded with innumerable requests that are practically impossible to handle by the system. The research article contributes to the research community by demonstrating step by step implementation and recovery of DDoS attack on SDN-based network.

The chapter is organized in the following manner. Section 2 describes the architecture of traditional networks followed by Sect. 3 of architecture of SDN. Section 4 talks about the controller and its architecture where the DDoS attack is launched. Section 5 talks about the simulation, traffic generation and analysis tools used during the experimentation. Section 6 provides the step by step procedure to implement the DDoS attack on SDN. Section 7 discusses the results and analysis followed by conclusion and references.

2 Traditional Network

In traditional networks, the data plane, control planes and management planes are coupled together in a device called switch. Figure 1 describes the switch architecture of traditional networks. It shows very clearly the coupling of data plane and forwarding plane. The control plane configures the node and determines the paths for the data flow. In this approach, once the data flow path is applied, any change at this level can be done only by reconfiguring the device. Frequent changes in the networks require reconfiguration of devices which is sometimes difficult to implement without disturbing the existing policies [8, 13].

As provided in Fig. 1, data plane consists of incoming and outgoing ports for sending and receiving data. The control plane provides the forwarding decisions fetched from the forwarding tables to the data plane. In this traditional approach, routers cannot choose the optimal path for data flow as it does not provide a global view of the network.

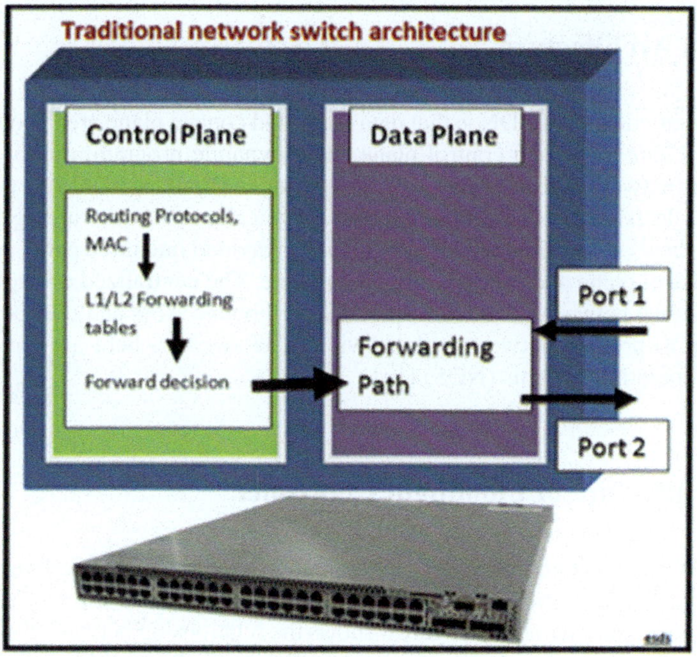

Fig. 1 Traditional network switch architecture

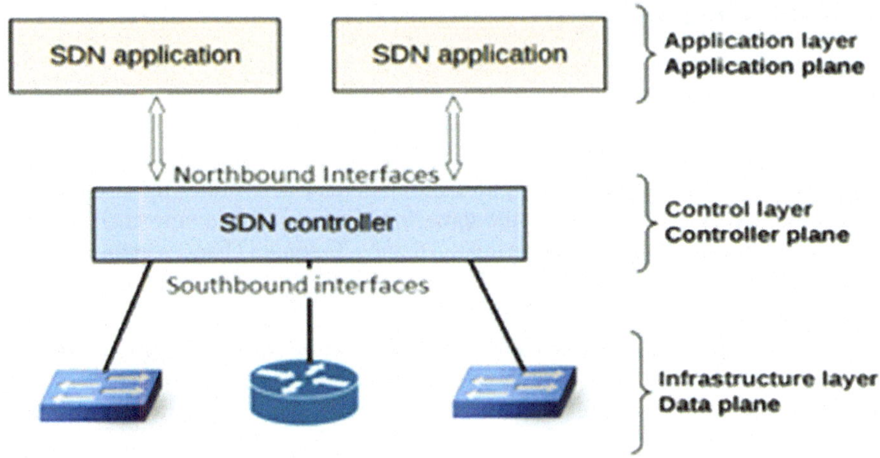

Fig. 2 An overview of SDN architecture

3 SDN Architecture

The peculiar concept of SDN is that data plane and control plane are decoupled. The decoupled data plane and control plane allow dynamic programming of data flow path in the network. While the control plane being the brain of SDN, it is performing the entire decision making and data plane is typically just a forwarding hardware. Figure 2 shows the architecture of SDN which is divided into three parts, application plane, control plane and infrastructure/data plane. The centralized controller works as an interface between the southbound and northbound APIs and thus provides the controller to provide network's global view. The switches here are controlled by network operating systems (NOS) [14].

4 Architecture of Floodlight Controller

Floodlight is two in one, controller plus collection of applications working on Floodlight Controller. Controller controls the network and applications provides common functionalities the user needs. Figure 3 shows the relationship between the controller and the built applications. The core architecture includes multiple modules such as topologies and path management, link discovery, routing, etc. which are internal dependant services. It also has core services such as provider, switch manager, performance monitor, message filters, etc. Floodlight Controller also provides utility services such as thread management, storage, test support, counters and servers of REST and Python. Module applications are supported for virtual network filtering, firewalls, forwarding functionalities, learning switch, static flow entry, hub, etc.

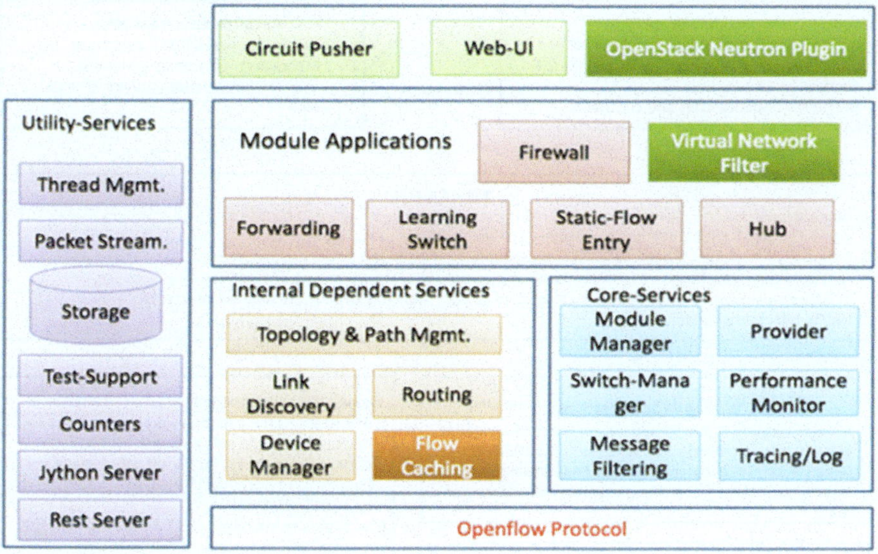

Fig. 3 An overview of the Floodlight Architecture

These modules communicate with Northbound APIs and Circuit Pusher, Web-UI, OpenStack Neutron Plugin, etc. Floodlight is a profound controller based on Java, with capability of upgradation and expansion which is the major reason for its selection for DDOS attack experiment by the authors. Figure 4 represents the step by step procedure to develop APIs for Java-based Floodlight Controller.

Developing modules in Floodlight is easy by converting the requirements into functionalities, events to response messages and utilities to procedures. Again, there is a rich set of existing modules of Floodlight that provides service through Java or REST APIs [4, 14].

5 Methodologies and Tools

In this section, we write about the simulators, emulators, controller and analyzing tools we have used in the experiment for data visualization to observe the data flowing through the network and for result analysis which is listed with its configuration versions in Table 1. As a test bed, a full-fledged virtual machine (VM) is developed with all the supporting tools and prerequisites installed before the experimentation begins. The base OS is Ubuntu with allocated resources listed in Table 1.

Mininet: Mininet is one of the emulators that are used to create virtual hosts and nodes customizable as per the requirement of experiment to study behaviour of each node under various controlled environments. The topologies are created using the

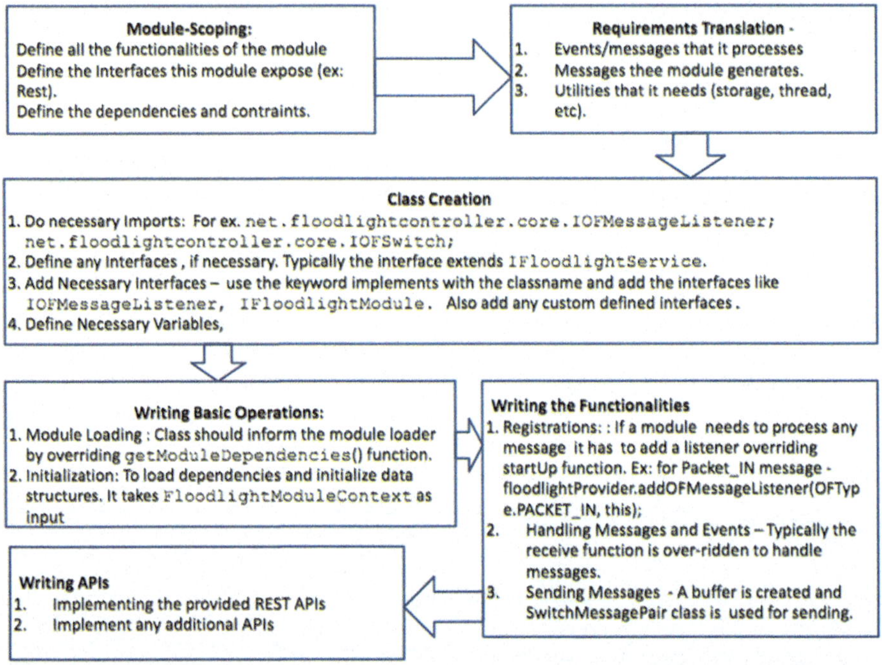

Fig. 4 Step by step procedure to develop REST or Java-based APIs for Floodlight Controller

Table 1 System configuration of virtual machine and simulation—emulation tools

Tools	Configuration
Operating System	Ubuntu 16.04 LTS
Processor	Intel Core i5
Memory	8 GB
Virtual Machine	VMWare 12
Mininet	2.2.0
IPERF	Version 3
XTERM	Default with Ubuntu
sFlow-RT	2.3
Gnuplot	5.2

Python script. One of the biggest benefits of using Mininet is that it can create very complex topologies to study and experiment without physically wiring up any [15]. **Python**: It is a general-purpose, high-level programing language that is popularly used due to its high flexibility and adaptability in network programming. With Python, we can perform various complex operations with lesser amount of coding. We used it to generate the topology in the Mininet [16].

sFlow-RT: The purpose of using sFlow-RT in this experiment is to give complete visibility into the network which also enables performance optimization along with defence against the security threats in presence of bridges. sFlow-RT is industry-standard analytical tool widely adopted by network vendors and end-users. Following step by step procedure guides you to install the tool for multiple purposes, whereas in this experiment, we use the sFlow-RT for research [17].

Step 1: Download the package for installation by typing the commands in the terminal.

wget https://inmon.com/products/sFlow-RT/sflow-rt.tar.gz

Step 2: Unzip the downloaded package and it will create a folder named sFlow-Rt

tar -xvzf sflow-rt.tar.gz

Step 3: Open the folder sFlow-RT folder through the terminal

cd sflow-rt

Step 4: Start the sFlow-RT by typing *./start.sh*

Step 5: Once started the sFlow-RT, it can be viewed in a browser on *localhost:8008*

Bridges: The visualization happens in sFlow-RT when bridges are implemented testing the diversified networking scenarios. Again, logical networking is provided using bridges on top of physical networking. Different nodes are to be connected logically from one to the other to analyze the flow of the traffic where virtual bridges play a vital role. The flow of the traffic from one node to the other happens only through bridges. These visualizations give a clear picture of the attack when it happens in sFlow-RT [18].

Xterm: It is a standard emulator for terminal functionalities which is used to create client–server connection between the two hosts. It is also used to create attack over server through client [19].

iPerf: The standard tool of network analysis is used in this experiment for tuning and traffic generation along with logging the network events [20].

Gnuplot: The purpose of using Gnuplot is to enhance the visualization of the obtained results in the form of data logs. Gnuplot supports filtering of each parameter using awk scripts and directly plots graphs of obtained results for result analysis. It also assures that no manipulation of data is done during the experimentation bringing trust between the authors and reviewers [21].

6 Implementation of DDoS Attack on SDN and Recovery

For the demonstration of the DDoS attack, a custom topology has to be generated in the Mininet using Python script. First, the traffic between two specific nodes is analyzed. After a particular time, a ping attack is generated from one node to the other and then later, the ping attack is terminated to recover the system for DDoS attack. As shown in Fig. 5, the topology has two intruder client who imposes DDoS attack on the software-defined network. From the available logs, graphs are plotted using Gnuplot to get the better understanding of the traffic before, during and after the DDoS attack. The detailed description of the whole procedure is as follows.

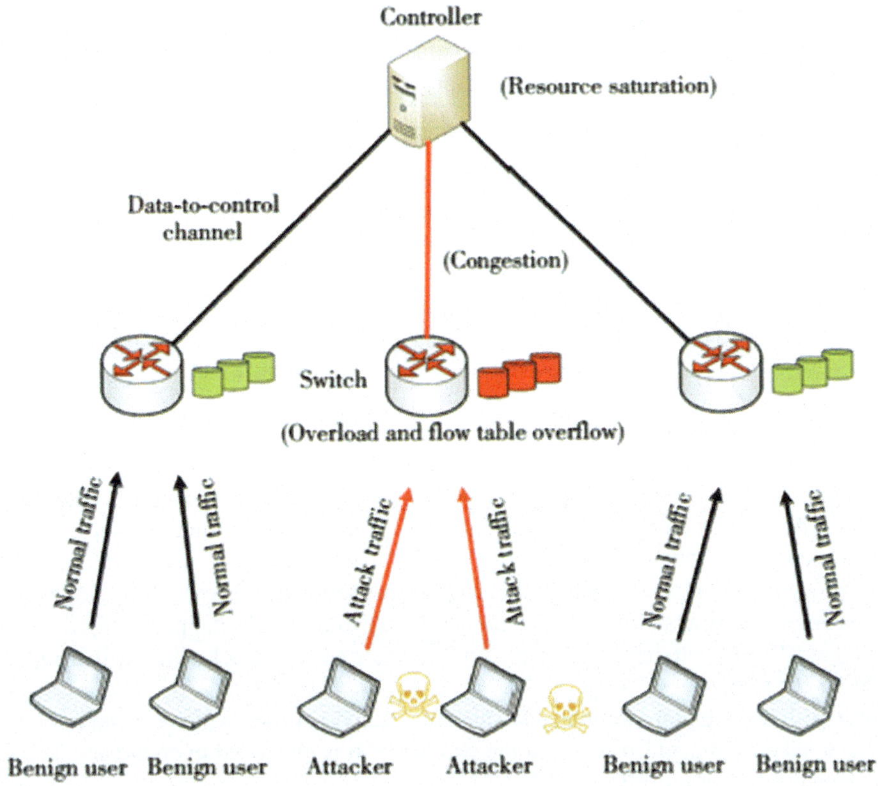

Fig. 5 DDOS attack on SDN controller, topology plan

Step 1: Open the Floodlight folder using the following commands.

$cd floodlight.

$java -jar target/floodlight.jar.

Now, open the browser and provide the following command to obtain GUI functionality of the controller. *localhost:8080/ui/index.html.*

Step 2:

Now, the topology needs to be created. The topology is made in Mininet with the help of the Python. The command that starts the Mininet is *mn*. Custom means the topology made by the user and not the default topology available in Mininet. The ip here is loopback (127.0.0.1), that is in the system itself and not in a remote location. The port 6653 is the port dedicated for the Floodlight Controller. Command:

sudo mn --custom ~/basic50.py --controller=remote,ip=127.0.0.1,port=6653 --topo mytopo

Type *pingall* on the Mininet and all the hosts and nodes will be available in the Floodlight which will be displayed in the topology located on the left-side panel. It is a linear topology having five switches and 50 hosts (ten each on every switch). Figure 6 shows the topology generated in Mininet on the Floodlight Controller.

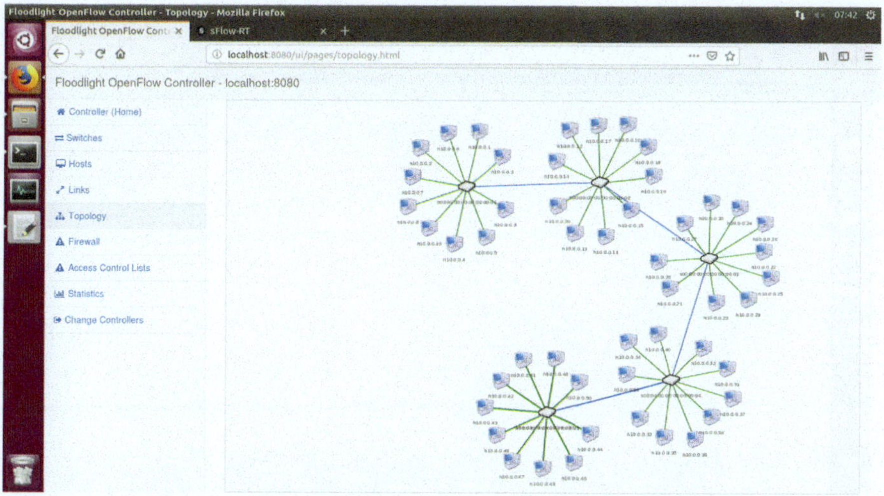

Fig. 6 Network topology with maximum five switches, each switch connected to 10 end systems

Step 3: Open the sFlow-RT and start it with the following commands:
$cd sflow-rt
$sudo ./start.sh

Now, open the browser and provide the following command to obtain GUI functionality. of the controller. *$localhost:8008/html/index.html.*

On the sFlow-RT, select the flows tab. On the name tab type flow. On the keys type ipsource, ipdestination, stack and on the value type bytes and then press submit. The purpose of this is to monitor the flow of the packets from the particular source to the destination. Figure 7 shows the interface of the sFlow-RT and its initial window appearance.

Step 4:
Type the following in the Mininet. There are 50 hosts in the topology. The analysis is made between h1 and h 50. *$xterm h1 h50.* Once again, *$xterm h1 h50.*

Here, four terminals are opened. Two for the host h1 and the other two for h50. In this one, pair is used for collecting the network traffic while in the other pair, the DDOS attack is launched.

Step 5:
The experiment starts from here. There are 300 s for which the experiment happens. Step 6 is set for 300 s. In the first 100 s, there is only Step 6 is executed. Next 100 s (101–200), Step 7 is executed where the sFlow-RT bridges are created and then Step 8 where the ping flood attack is launched on the controller. After 200 s (means 100 + 100 s), Step 10 is executed where the DDOS attack is brought down by closing the sFlow-RT bridges and ping flood process. From 201–300 s, the readings are for the effect of DDOS attack on the network traffic. In this way network traffic before DDOS attack (1–100 s), during (101–200 s) DDOS, and after (201–300 s) DDOS can be analyzed.

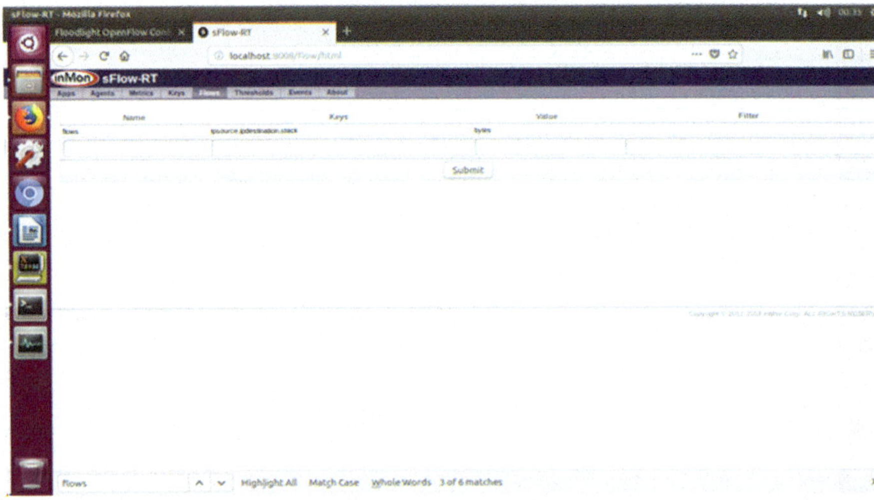

Fig. 7 sFlow-RT implementation window once it starts running

Step 6:
On the left-side, terminals (Fig. 8) are used for collecting the network traffic reading. Iperf3 is used to collect the network traffic.

On the left-side top terminal node h1 type *iperf3 -s -f G -p 5566 -i 1 > server.txt*

On the left-side bottom terminal node h50 type *iperf3 -c 10.0.0.1 -f G -p 5566 -t 300 > client.txt.*

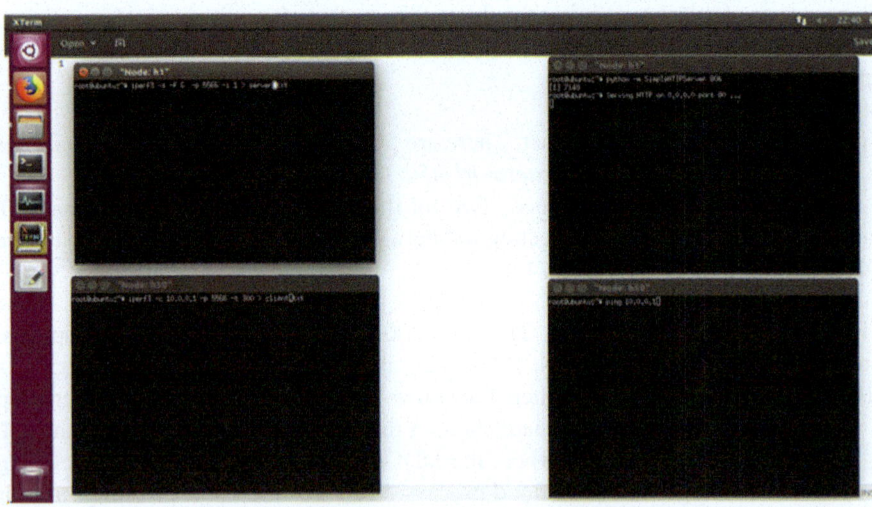

Fig. 8 XTerm windows while generating traffic between client and server before injecting DDoS attack

The above two commands run on two left terminals collect the network traffic in every one second for a time period of 300 s in total.

In the above commands, *-s* stands for server, *-c* stands for client, *-f* is for format, *G* for GBytes, *-p* means port and *-t* is time.

Step 7:

The sFlow-RT needs to be configured well such that it can display the analysis part well. There are bridges that need to be created between the switches. Here, 'ovs' stands for Open vSwitch. The operation happens here is the configuration of bridge s1 to s5 to send sFlow records to a collector on 127.0.0.1 at port 6343, using eth0´s IP address as the source. The port for the sFlow-RT to receive is 6343.

Execute these commands on a new terminal where each command is for individual five configured switches.

sudo ovs-vsctl -- --id=@sflow create sflow agent=eth0 target=\"127.0.0.1:6343\" sampling=10 polling=20 -- -- set bridge s1 sflow=@sflow

sudo ovs-vsctl -- --id=@sflow create sflow agent=eth0 target=\"127.0.0.1:6343\" sampling=10 polling=20 -- -- set bridge s2 sflow=@sflow

sudo ovs-vsctl -- --id=@sflow create sflow agent=eth0 target=\"127.0.0.1:6343\" sampling=10 polling=20 -- -- set bridge s3 sflow=@sflow

sudo ovs-vsctl -- --id=@sflow create sflow agent=eth0 target=\"127.0.0.1:6343\" sampling=10 polling=20 -- -- set bridge s4 sflow=@sflow

sudo ovs-vsctl -- --id=@sflow create sflow agent=eth0 target=\"127.0.0.1:6343\" sampling=10 polling=20 -- -- set bridge s5 sflow=@sflow

Step 8:

Out of the four xterms opened, in the right-side top node h1 type the command *$python -m SimpleHTTPServer 80&*

The above command makes node h1 as the server. In the right-side bottom node h50 type the command *$ping -f 10.0.0.1*

10.0.0.1 corresponds to the inet address of h1 and *-f* means flooding. The above commands means that there is ping flood attack is launched from h50 to h1. Figure 8 shows the four different xterm windows. Left side two are corresponding to the Step 6 and right side are corresponding to Step 8. The left ones are used to collect the traffic details and right ones are for the DDoS attack.

Step 9:

The analysis can be viewed with the help of the sFlow-RT. In Step 3, the flows tab is configured. The source and destination traffic can be analyzed well in the interface. The packets are sent in bytes. Now, open the agents tab and select the 127.0.0.1. It will provide the statistics that are discussed in the result analysis section.

Recovering from DDoS Attack:

Step 10:

In order to stop the DDOS attack, we need to stop the process. See in which process the ping is happening and then kill that process. Type top in a new terminal. In the processes, see id for the ping and then press q. Figure 9 shows the termination of the DDoS attack by terminating the ping command that is the cause in this simulation.

Type *sudo kill <process id>* (enter the process id. Here, ping Id is 12049 as seen in the figure.)

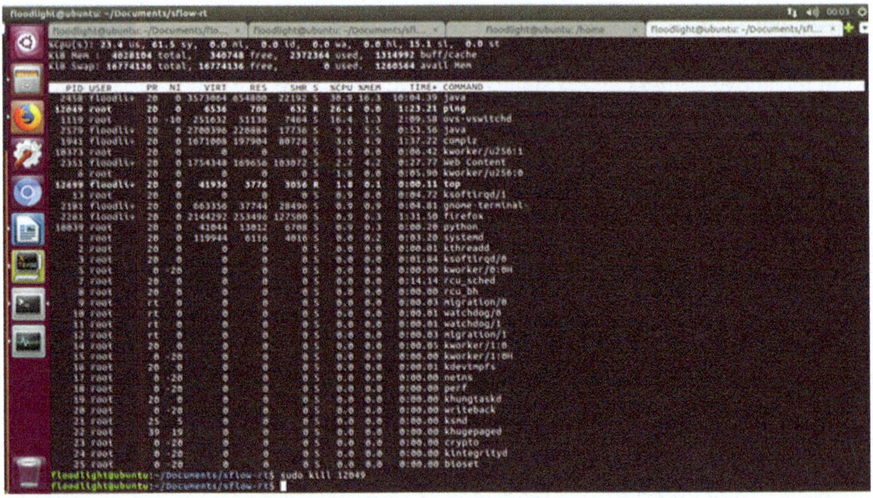

Fig. 9 Process to overcome the DDoS attack while monitoring the resource consumption of the processes

Once the process is killed, the DDOS comes down. Figure 10 shows how the fall happens from a state of high traffic to low traffic. The moment the cause is terminated, the attack comes to an end. The direct fall is the point where the attack has come to an end.

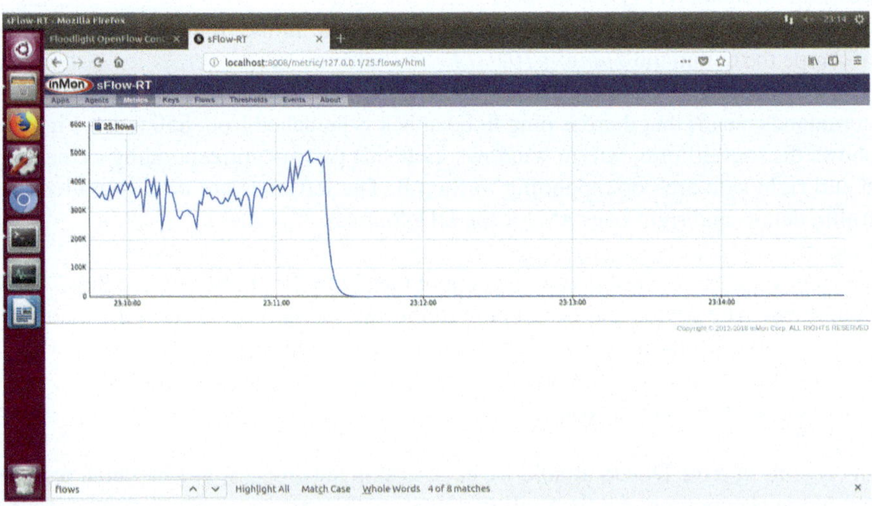

Fig. 10 Visualization of implementing ping attack and after resolving attack in sFlow-RT

7 Result Analysis

This section analyzes the results obtained in the form of graph after logging the events before and after the DDoS attack implemented on SDN. In this experiment, the authors have made attempt to explore the DDoS attack situation on Floodlight Controller and test its resilience against the attacks.

The major issue we faced during the experimentation with Floodlight while implementing DDoS attack was that when ping attack is implemented, it makes the system so slow that hardly recovery becomes possible to uplift the system back to its originality. We brought solution to this issue by reducing the flows demanding internet which can be achieved by requesting network users to shut down their systems. Another issue we faced was of a large variation observed during each simulation run of the experiment for both the scenarios irrespective of the presence or absence of bridges which was resolved by running minimum 25 runs of each experiment and taking the most repeated records in account.

The first target for any intruder to implement DDoS attack over the network is the switches. So, taking the realistic approach, we have implemented attacks on two different scenarios. One scenario is with three switches where the small scale business is represented by this topology. The second scenario is of the five switches which represent cascading of switched network with one router representing one branch of the large scale networks. The small sub-networks which are a part of the large scale networks are the gateways for intruders to impose attack on the network of large scale businesses located far apart in different continents. The control over the attack has to be obtained no sooner the attack is encountered before it infects the entire network. Thus, we have limited the study till small scale networks and simulated with maximum of five switched networks.

The graph represented in Fig. 11 represents throughput logged from 0 to 300 s where the attack is launched at 100 s and recovered at 200 s ending the experiment at 300 s. The same simulation setup is followed for upcoming all the scenarios which can be observed in upcoming graphs. It can be observed from the figure that from 0 to 100 s, there are hardly any events of packet drop. Again, the stable data transmission occurs until 98 s. The moment DDoS attack is imposed, the throughput reduces to 0 with minor variations and hardly any packet is delivered after 100 s. No sooner at 200 s the DDoS attack is recovered, the throughput again reaches at its best between 2 to 2.5 Gbps and remains stable until the simulation ends. Here, topology has presence of bridge which makes the graphs stable by logically managing the flows, an added advantage over the bottleneck networks.

Graph provided in Fig. 12 shows the throughput observed in absence of bridge over sFlow-RT for three switches. There is a recognizable behavioural difference between throughput before applying and after recovery of DDoS attack. The packet drops observed before applying the attack is far more in comparison with the packet drops observed after the recovery from DDoS attack. The reason behind it may be the filled entries of switch tables. There is an impact on network because of absence of bridge which can be observed as the data is not managed logically, network is

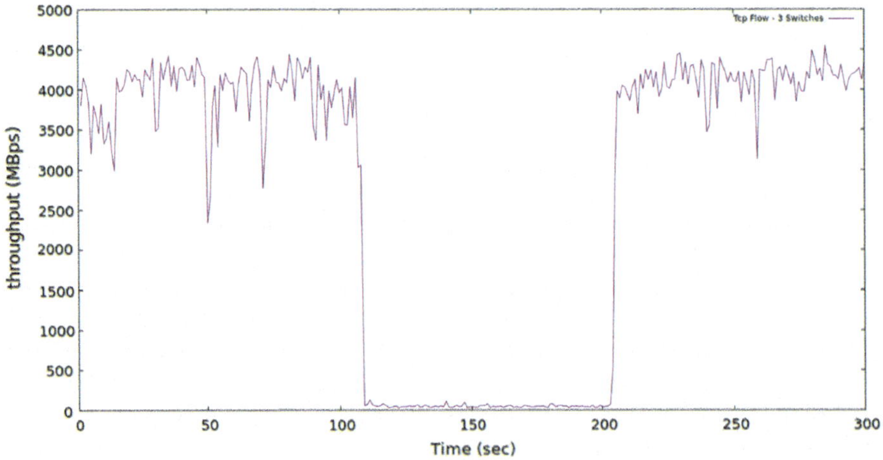

Fig. 11 Throughput before the DDoS attack, during the attack and after the recovery for THREE nodes WITH sFlow-RT bridge implementation

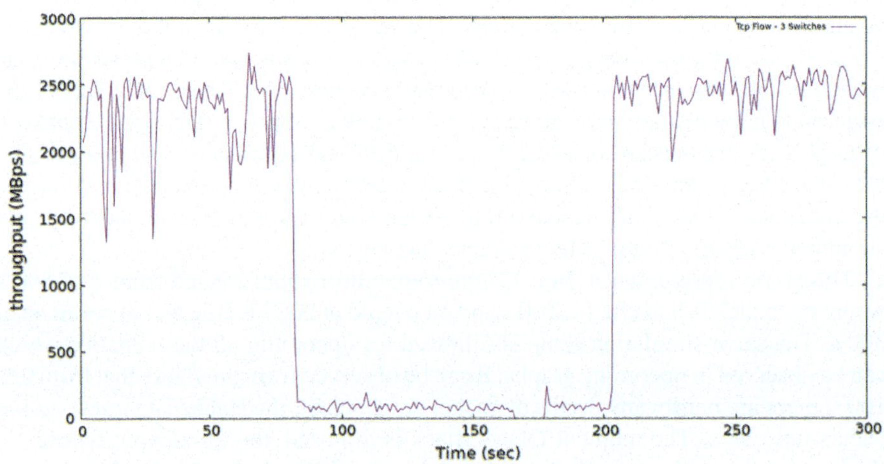

Fig. 12 Throughput before the DDoS attack, during the attack and after the recovery for THREE nodes WITHOUT sFlow-RT bridge implementation

not able to handle the burst flow situations and events of packet drops are more in comparison with Fig. 11.

As shown in Fig. 13, we observe the throughput over five nodes before and after DDoS attack imposed in presence of the sFlow-RT bridges which provides additional logical network distribution support. It can be observed from the graph that as the number of switches increased to five along with additional 20 connected end host, the network seems to be quietly unstable to handle the demand of packet delivery and few drop events were observed before the attack is imposed. No sooner the attack is

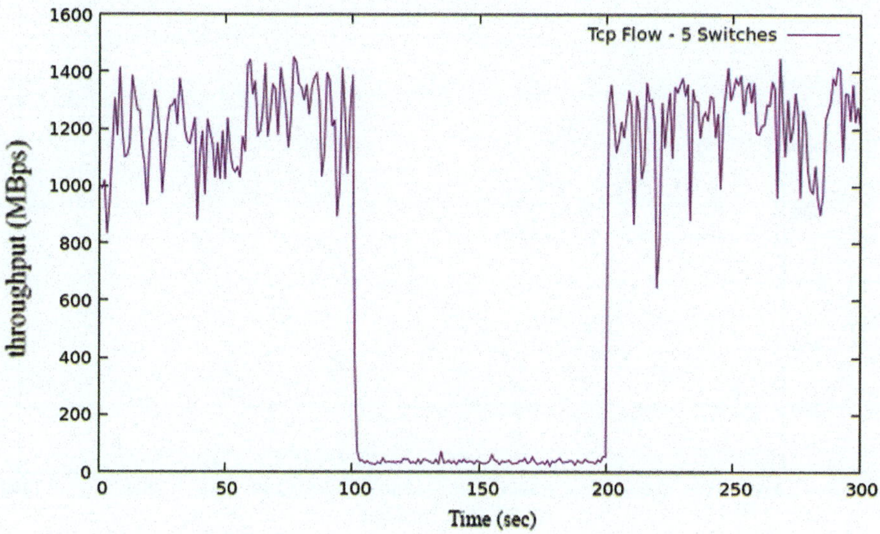

Fig. 13 Throughput before the DDoS attack, during the attack and after the recovery for FIVE nodes WITH sFlow-RT bridge implementation

imposed, packets received and sent reached to count zero until system recovers. Even in presence of the bridge, the stability of throughput is not observed after the recovery from DDoS attack. The reason may be the presence of burst traffic condition along with the bottleneck situation due to additional 20 hosts actively creating network congestion.

As shown in Fig. 14, graph represents throughput before the DDoS attack, during the attack and after the recovery for five nodes without sFlow-RT bridge implementation. The reason behind stability after recovery from DDoS may be because the switch table is filled with required network information, and thus, initial broadcasting is not required when system is recovered.

The moment system recovers at 200 s, network becomes stable and drop events are reduced drastically after the recovery and keep improving its stability till the end of simulation. The reason behind this behaviour is that for large scale networks, logical network management through bridges may prove to be heavy and not recommended for large scale networks, as it becomes overhead for the network.

The graph provided in Fig. 15 observes the congestion window for the TCP flows transferring through the network of three switches in presence of sFlow-RT bridges. It can be observed that the congestion window remains in the range of 0.5–1.5 Mbps before the attack is imposed. Window size reduces drastically when the attack is imposed as the network appears to be congested due to DDoS attack. No sooner the network is recovered, a boost is observed in the beginning as the congestion is observed to be zero which takes few RTTs to update. Later, it becomes stable but stays high in the range of 0.5–2 Mbps.

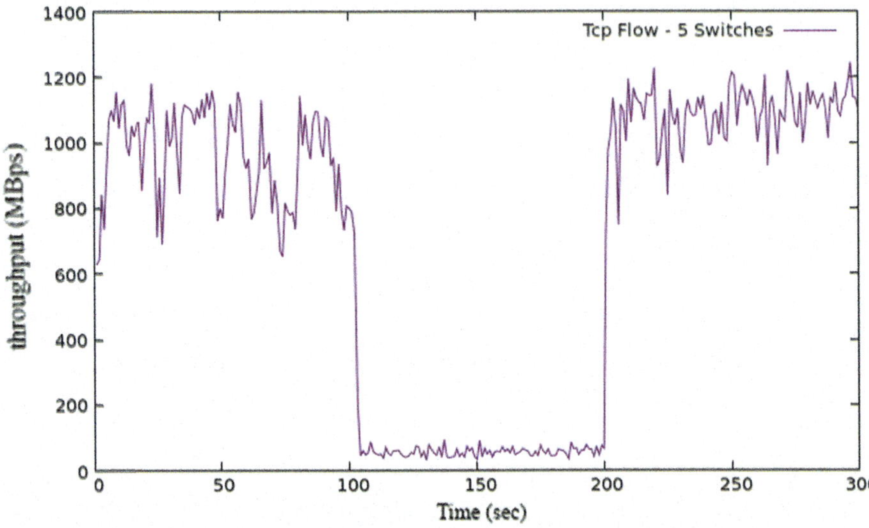

Fig. 14 Throughput before the DDoS attack, during the attack and after the recovery for FIVE nodes WITHOUT sFlow-RT bridge implementation

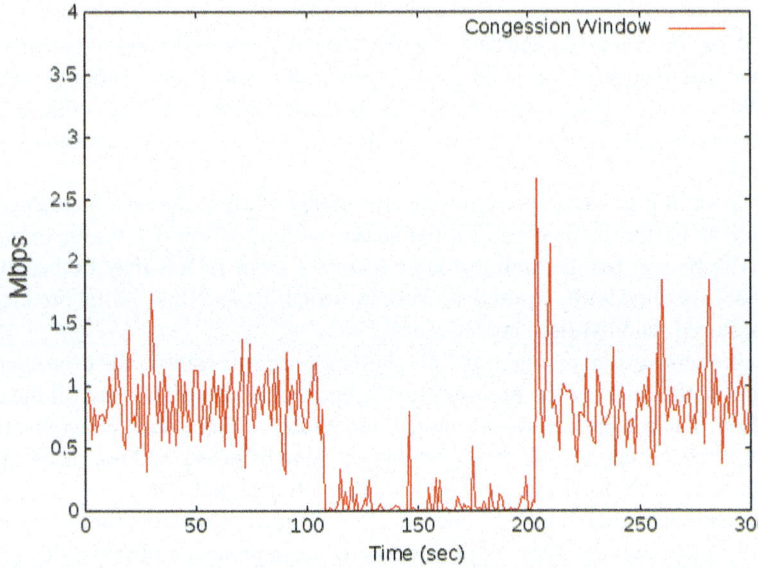

Fig. 15 Congestion window during the DDoS attack and after the recovery WITH bridge on THREE switches

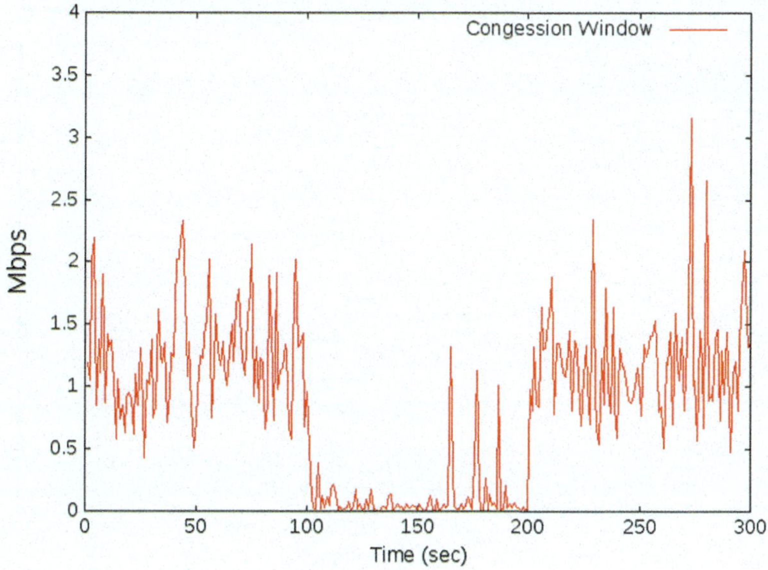

Fig. 16 Congestion window during the DDoS attack and after the recovery WITHOUT bridge on THREE switches

Figure 16 represents congestion window of the three-switch scenario observed in absence of the bridge. It can be seen that the congestion window stays unstable throughout the experiment which proves that there is a positive impact of presence of bridge and its absence may result to the unstable behaviour of network parameters as shown in Fig. 16. Lot of deeps and steep are observed throughout the experiment.

Figure 17 represents congestion window of the scenario with five switches and 20 additional nodes in presence of bridge. It can be observed that no sooner the scale of network increases to five switches; it behaves unstable and goes worse even after the recovery from DDoS attack. The range of congestion window was 0.5–2.2 Mbps before the attack was imposed which increases to the range of 0.2–3 Mbps. It is clear from the graph that bridges are not recommended for congestion control over SDNs.

The graph represented in Fig. 18 shows the positive impact of absence of bridge on the congestion window as far as large scale networks are concerned. We can observe from the graph which is of five switches and 20 additional networks in absence of bridge is far stable in comparison with the graph provided in Fig. 17. This proves that for congestion control stability, presence of bridge is a negative factor and must be avoided.

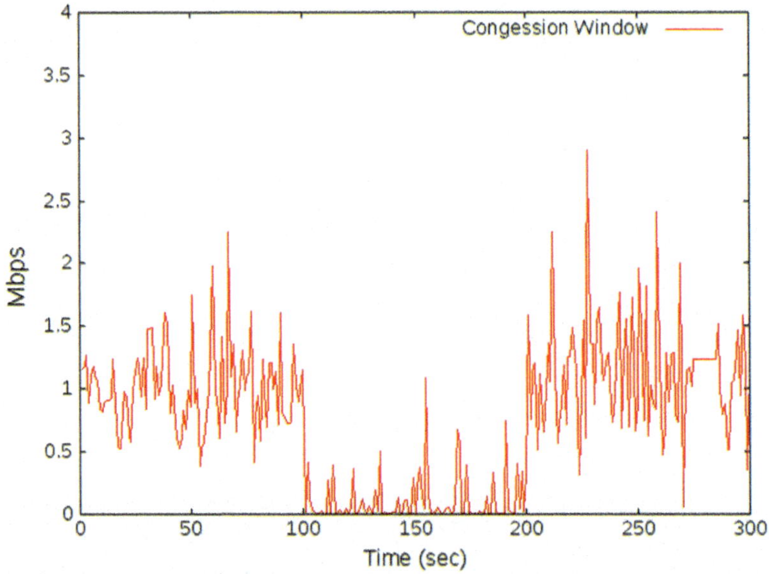

Fig. 17 Congestion window during the DDoS attack and after the recovery WITH bridge on FIVE switches

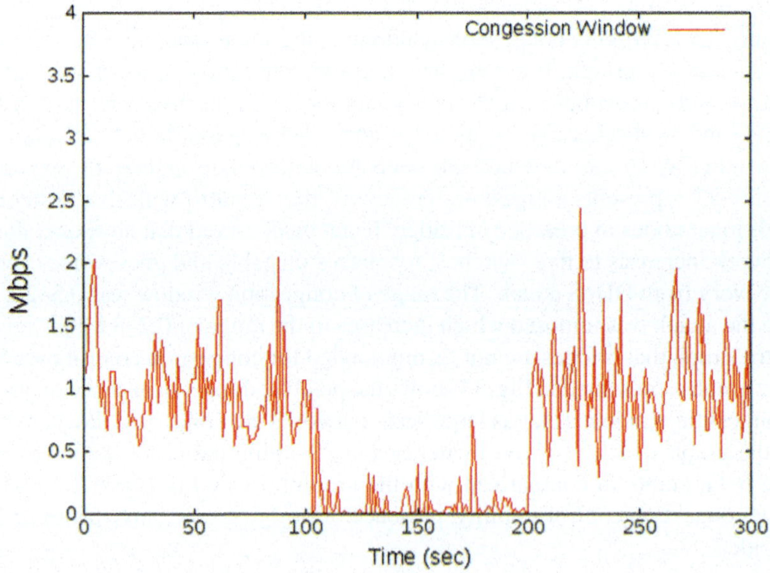

Fig. 18 Congestion window during the DDoS attack and after the recovery WITHOUT bridge on FIVE switches

8 Conclusion

Thus, with this experiment, authors have demonstrated the step by step procedure to implement DDoS attack on Floodlight Controller and its recovery techniques. During the experiment of security attack, authors also throw light upon the existing data visualization techniques available in the area of SDN especially focusing on parameters of analysis of DDoS attack on SDN with clarity on its usage, features, applicability and scopes for its adaptabilities in the world of networks which is the future of the booming networking innovations. The readers of this experiment will benefit whether a beginner or an expert in the domain of networks as its sections and subsections show clearly the experimental steps to implement DDoS attack on SDN and further provides solution to overcome the attack. The newbie in the area of SDN can recreate the experiment with modifications as per their requirements which will give them a practical exposure of the implementation of concepts of security over SDN with confidence and in-depth wisdom through visualization of obtained resultant data plotting as graphs. The researchers in the area of security and SDN can expand the work done in this experiment by adding complexities of networking scenario and security aspects with confidence on its feasibility as a part of work is clearly demonstrated through this experiment. This chapter opens doors to research opportunities in multiple dimensions in the area of security and software-defined networks with multiple options of data visualizations.

References

1. Asadollahi, S., & Goswami, B. (2017). Software defined network, controller comparison. *IJIRCCE, 5*(2), 211–217.
2. Das, S., Goswami, B., & Asadollahi, S. (2017). Investigating software-defined network and networks-function virtualization for emergent network-oriented services. *IJIRCCE, 5*(2), 201–205.
3. Asadollahi, S., Goswami, B., & Sameer, M. (2018). Ryu controller's scalability experiment on software defined networks. In *Proceedings of IEEE International Conference on Current Trends in Advanced Computing (ICCTAC)* (pp. 1–5). IEEE, Bangalore, India.
4. Asadollahi, S., & Goswami, B. (2017). Experimenting with scalability of floodlight controller in software defined networks. In *International Conference on Electrical, Electronics, Communication, Computer, and Optimization Techniques (ICEECCOT)* (pp. 1–5). IEEE, Mysore, India.
5. Goswami, B., & Asadollahi, S. (2017). Implementation of SDN using OpenDayLight controller. *IJIRCCE, 5*(2), 218–227.
6. Manuel, T., & Goswami, B. H. (2019). Experimenting with scalability of beacon controller in software defined network. *International Journal of Recent Technology and Engineering (IJRTE), 7*(5S2), 550–555.
7. Sameer, M., & Goswami, B. (2018). Experimenting with ONOS scalability on software defined network. *Journal of Advanced Research in Dynamical & Control Systems, 10*(14-Special Issue), 1820–1830.

8. Goswami, B., & Asadollahi, S. S. (2018). Enhancement of LAN infrastructure performance for data center in presence of network security. In: Lobiyal, D., Mansotra, V., & Singh, U. (Eds.), *Next-generation networks. Advances in intelligent systems and computing* (vol. 638). Singapore: Springer.
9. Asadollahi, S., & Goswami, B. (2017). Revolution in existing network under the influence of software defined network. In *Proceedings of the 11th INDIACom* (pp. 1012–1017). IEEE, New Delhi, India.
10. Asadollahi, S., & Goswami, B. (2016). Key establishment technique for secure diversified wireless network. In *ICCSNIT—2016*, Pattaya, Thailand. Open Access.
11. Goswami, B. (2012). *Study and analysis of symmetric key-cryptograph DES, data encryption standard*. SWP-2012, Rajkot. India. MHRD Sponsored. India.
12. Hameed, S. S., & Goswami, B. (2018). SMX algorithm: A novel approach to avalanche effect on advanced encryption standard AES. In *Proceedings of the 12th INDIACom* (pp. 727–232). IEEE, New Delhi, India.
13. Gosai, M., Goswami, B., & Kar, U. (2014). Experimental based performance testing of different TCP protocol variants in comparison of RCP+ over hybrid network scenario. *International Journal of Innovations & Advancement in Computer Science (IJIACS), 3*(2), 31–37.
14. Asadollahi, S., & Goswami, B. (2017). Scalability of software defined network on floodlight controller using OFNet. In *International Conference on Electrical, Electronics, Communication, Computer, and Optimization Techniques (ICEECCOT)* (pp. 1–5). IEEE, Mysore, India.
15. De Oliveira, R. L. S., et al. (2014). Using mininet for emulation and prototyping software-defined networks. In *2014 IEEE Colombian Conference on Communications and Computing (COLCOM)*. IEEE.
16. Fettig, A., & Lefkowitz, G. (2005). *Twisted network programming essentials*. O'Reilly Media, Inc.
17. Rehman, S. U., Song, W.-C., & Kang, M. (2014). Network-wide traffic visibility in OF@ TEIN SDN testbed using sFlow. In *The 16th Asia-Pacific Network Operations and Management Symposium*. IEEE.
18. Visitsathapong, C. (2014). *Path switching delay measurements in software-defined networks*. Diss.
19. Xterm: Emulator. Available at https://invisible-island.net/xterm/. Last accessed on September 2018.
20. IPERF: Networks tool. Available at https://iperf.fr/. Last accessed on September 2018.
21. Gnuplot: Graph tool. Available at http://www.gnuplot.info/. Last accessed on September 2018.

Data Visualization of Software-Defined Networks During Load Balancing Experiment Using Floodlight Controller

Mohammed Asif Khan, Bhargavi Goswami and Saleh Asadollahi

Abstract With the growing impact of globalization, large-scale networks are now not countable with the tip of fingers as businesses have grown drastically and every day more and more organization are stepping into the lobby of large-scale networks. While the entire network is allocated with a diverse set of resources, it is necessary to address the balance of the usage of Internet resources to avoid long waiting queues and congestions over certain hotlines of communication channels that have always been the focus for the entire market. As the traditional network has evolved to software defined-networks (SDN), for addressing load balancing issue over SDN, we as researchers have come forward to provide the solution to the problem by means of this chapter. Here, researchers have discussed and demonstrated the implementation of the unique technique of load balancing on software-defined networks on Floodlight Controller over mininet simulator with the focus on Data Visualization while performing the experiment.

Keywords SDN · Mininet · OpenFlow · Load balancing · IPerf · Floodlight · RestAPI · Gnuplot

1 Introduction

By using programmable controllers, the technology of SDN—software-defined networks has generated a separation wall between the logic of data and control plane. In SDN, the centralized controller has a global view of networks and it is aware of every single event affecting the network which makes it capable of managing the network operations efficiently. A whole set of variety of operations is supported by the control plane of controllers to become the backbone of high-performing networks. But, if

M. A. Khan
Department of Computer Science, CHRIST, Bangalore, India

B. Goswami (✉)
School of Electrical Engineering and Computer Science, QUT, Brisbane, Australia

S. Asadollahi
ITTECS, Brisbane, Australia

S. M. Anouncia et al. (eds.), *Data Visualization*,
https://doi.org/10.1007/978-981-15-2282-6_9

we talk about the large-scale networks, more than one controller plays the role in managing the distributed network physically along with being a part of centralized controlled networks logically. However, the growing requirement of shift from traditional network to software-defined networks leads to the situations of congestion and clustering that needs to be addressed before the breakdown situation is created. To address this issue, best suitable solution seems to be monitoring the networks for clusters and congested areas. Further, there is a clear requirement of implementation of load balancing techniques no sooner the clusters and congested areas are observed in the network. Load balancing is implemented in two states, (1) link load balancing and (2) controller load balancing. To provide better performance, the trade-off link and controller load balancing is performed resulting to the expected high performance with reduced load on the network. Again, there is a significant impact of load balancing on the resource utilization and energy consumption as far as actions are taken on time before the network trumbles down.

The authors of [1] have shown the HTTP request redirected to predefined servers and demonstrates the load balancing using round-robin algorithm for large number of connections distributed uniformly in the network. Authors have not shown the analysis of the network in comparison of situation before and after the implementation of load balancing techniques.

Other than protocol and specialized softwares, load balancing can also be obtained using specialized gateways over LANs. This is demonstrated by the authors of [2] where the analysis is not done as per the expectations of the research standards without considering the noise.

The same load balancing is under progress to be implemented upon the multiple controllers which has also been experimented for scalability such as Ryu [3], Floodlight [4, 5], OpenDaylight [6, 7], Beacon [8], Onos [9], Cisco LANs [10], and wide variety of real-time networks [11, 12] to compare the performance in presence of wide variety of SDN controllers. In this chapter, our focus is limited to Floodlight Controller as it has not been implemented so far to our knowledge.

The chapter is distributed in the following sections. Section 2 describes the step-by-step implementation of Floodlight Controller. Section 3 demonstrates development of SDN-based scenario. Section 4 demonstrates implementation of load balancing on SDN, Sect. 5 discusses the experimental issues and its resolution, Sect. 6 explains how the complex graphs are generated using scripts, and Sect. 7 is discussion on performance analysis followed by Conclusion and References.

2 Implementing Floodlight Controller on SDN

The objective is to perform load balancing over SDN and at the same time minimizing the delay that happens in a data communication over the network, i.e., Latency. Here, Dijkstra's algorithm is used to find multiple paths of same length which enables us to reduce the search to a small region in the fat tree topology. Specific rules might need to be pushed to get a proper load balancing output. Currently, the load balancer

Table 1 Configuration of platform and tools used during experimentation

Oracle VM virtual box manager	5.2.24 r128163 (Qt5.6.2)
Ubuntu	14.04 64-bit
Mininet	2.2.1
Iperf	2.0.5
Gnuplot	4.6
Openflow	1.4
Processor	2 CPUs
Base memory	2.5 MB
Floodlight	v2.1
Wireshark	1.12.1

program simply finds the path with least load and forward traffic on that path. To implement the load balancing on SDN network using Floodlight Controller, multiple simulators are installed on the Ubuntu platform over a virtual machine along with some traffic generation and analysis tools. The specification the platform along with the simulation tools is provided with version in Table 1.

It is observed from the table that over virtual machine, Ubuntu platform is created that has two CPUs with 2.5 MB base memory, on which some research tools such as mininet, OpenFlow, Floodlight Controller, iPerf, Wireshark, Gnuplot are used.

3 Developing SDN-Based Scenario

A step-by-step guide for performing the experiment on Floodlight Controller [13] using mininet is provided below. This section explains the steps to be followed before load balancing techniques that are implemented to set up the environment. The first requirement is to generate the software-defined network using Floodlight Controller. Figure 1 represents the networking scenario generated for the experiment where top two nodes represent directly connected programmable switches to the Floodlight Controller which takes the commands from controller directly. Remaining non-terminal nodes are intermediate switches and terminal nodes are hosts.

Step 1: The first step is to run the floodlight controller from the floodlight folder in a new terminal. Please note that throughout the experiment, the used nomenclature is specific to our system which is a variable terminology specific to the one's system specification. The name of the folder is floodlight. Command: cd floodlight. Now, running the floodlight controller using the command: java -jar target/floodlight.jar. This command will start the floodlight controller. To avoid any packet loss in the network, before proceeding to the next step ensuring the controller is sending the LLDP packets from all the enabled ports.

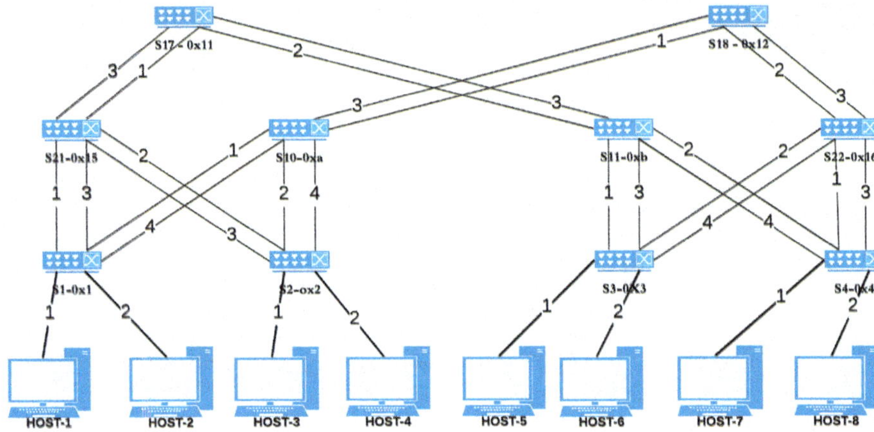

Fig. 1 Networking scenario developed on software-defined network using Floodlight Controller

Step 2: This step is to run the mininet [14] mesh topology script, by specifying the topology name, ip address of the system, and the port number with the following command: sudo mn --custom topology.py --topo mytopo --controller=remote, ip=127.0.0.1, port=653. By default, OVS switch [15] is used by the mininet simulator. Once this command is executed successfully, check the connectivity between all the hosts using the mininet command: Pingall. Performing pingall as long as there is no packet loss in the network. Note: topology.py is a python [16] script as given in Fig. 2, written to overwrite the default topology of simulator so that customized complex topology can be generated and experimented. Figure 3 shows the visual on Topology Tab on Floodlight Controller which is same as provided in Fig. 1.

Step 3: After successful execution of the above steps, the client and server are defined among the available hosts using xterm [17]. Before defining this, check the communication of the hosts over the network with other hosts. The command to perform this is: xterm h1 h1. In the first console of h1 type: ping 10.0.0.3 and in the second console of h1 type: ping 10.0.0.8. When this command is executed successfully, Host-1 is able to communicate with Host-3 and Host-8 over the network.

Step 4: In this step of the experiment, we define one client and two servers from the developed network. The command that is used for this is: xterm h1 h3 h8. This command will open three consoles h1, h3, and h8. To check the configuration details on each of these consoles, type the command: ifconfig

Step 5: Now, the step by step procedure followed for generating the traffic between client and the servers and recording the events using iperf tool [18]. (a) In the h3 console, type the command: iperf -s -p 6653 -i 1 > result-H3. Here, defining the Host-3 with ip address 10.0.0.3 as the server. Where, "result-H3" is the filename to store the events occurring between Host-1 and Host-3. After this command is executed, the server starts and waits for the client request on the network. Now in the h1 console, type the command: iperf -c 10.0.0.3 -p 6653 -t 100. Here, for generating the traffic at the client side 10.0.0.1, providing the server ip address 10.0.0.3 and the

```
from mininet.node import CPULimitedHost, Host, Node
from mininet.node import OVSKernelSwitch
from mininet.topo import Topo

class fatTreeTopo(Topo):

  "Fat Tree Topology"

  def __init__(self):
    "Create Fat tree Topology"

    Topo.__init__(self)

    #Add hosts
    h7 = self.addHost('h7', cls=Host, ip='10.0.0.7', defaultRoute=None)
    h8 = self.addHost('h8', cls=Host, ip='10.0.0.8', defaultRoute=None)
    h1 = self.addHost('h1', cls=Host, ip='10.0.0.1', defaultRoute=None)
    h2 = self.addHost('h2', cls=Host, ip='10.0.0.2', defaultRoute=None)
    h4 = self.addHost('h4', cls=Host, ip='10.0.0.4', defaultRoute=None)
    h3 = self.addHost('h3', cls=Host, ip='10.0.0.3', defaultRoute=None)
    h5 = self.addHost('h5', cls=Host, ip='10.0.0.5', defaultRoute=None)
    h6 = self.addHost('h6', cls=Host, ip='10.0.0.6', defaultRoute=None)

    #Add switches
    s10 = self.addSwitch('s10', cls=OVSKernelSwitch)
    s3 = self.addSwitch('s3', cls=OVSKernelSwitch)
    s17 = self.addSwitch('s17', cls=OVSKernelSwitch)
    s4 = self.addSwitch('s4', cls=OVSKernelSwitch)
    s18 = self.addSwitch('s18', cls=OVSKernelSwitch)
    s1 = self.addSwitch('s1', cls=OVSKernelSwitch)
    s11 = self.addSwitch('s11', cls=OVSKernelSwitch)
    s21 = self.addSwitch('s21', cls=OVSKernelSwitch)
    s22 = self.addSwitch('s22', cls=OVSKernelSwitch)
    s2 = self.addSwitch('s2', cls=OVSKernelSwitch)

    #Add links
    self.addLink(h1, s1)
    self.addLink(h2, s1)
    self.addLink(h3, s2)
    self.addLink(h4, s2)
    self.addLink(h5, s3)
    self.addLink(h6, s3)
    self.addLink(h7, s4)
    self.addLink(h8, s4)
    self.addLink(s1, s21)
    self.addLink(s21, s2)
    self.addLink(s1, s10)
    self.addLink(s2, s10)
    self.addLink(s3, s11)
    self.addLink(s4, s22)
    self.addLink(s11, s4)
    self.addLink(s3, s22)
    self.addLink(s21, s17)
    self.addLink(s11, s17)
    self.addLink(s10, s18)
    self.addLink(s22, s18)

topos = { 'mytopo': (lambda: fatTreeTopo() ) }
```

Fig. 2 Code snippet for the network

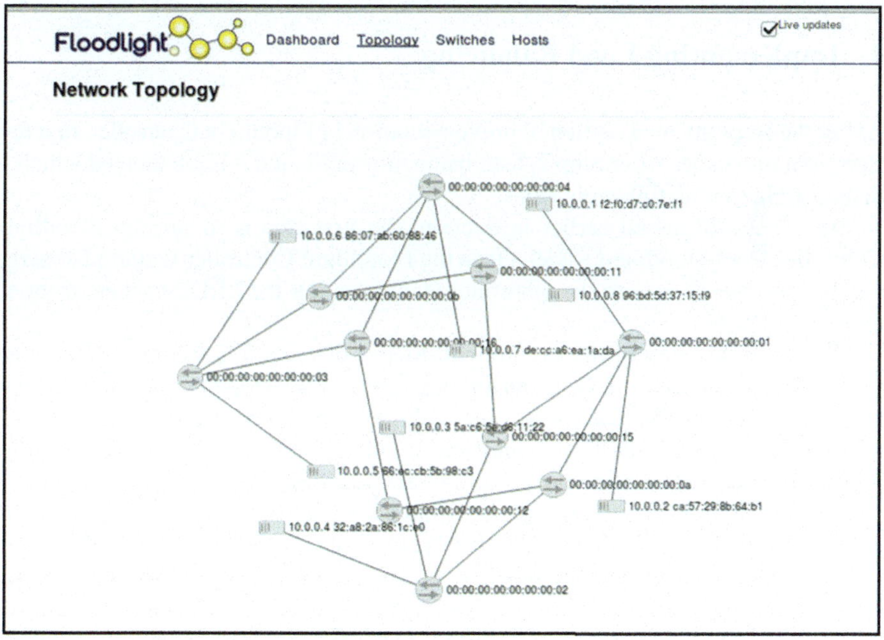

Fig. 3 Network topology generated and visible on topology tab of Floodlight Controller

port number on which the server is listening is 6653. 100 represents time in seconds. (b) Before proceeding with this step, stop the Host-3 server. In the h8 console, type the command: iperf -s -p 6653 -i 1 > result-H8. Now, define the Host-8 with ip address 10.0.0.8 as the server, where "result-H8" is the filename to store the events occurring between Host-1 and Host-8. After this command is executed, the server starts and waits for the client request on the network. Now, in the h1 console, type the command: iperf -c 10.0.0.8 -p 6653 -t 100. Here, for generating the traffic at the client side 10.0.0.1, provide the server ip address 10.0.0.8 and the port number on which the server is listening is 6653.

Step 6: Next step is to filter specific results that are obtained from the experiment. From the generated result files in Step 5, further the experiment is continued with these files. For filtering the results, grep and awk commands are used for two files as follows: (a) Command: cat result-H3 | head -106 | grep sec | awk '{print $3,$5}' > output_H1-H3. Here, "output_H1-H3" is the name of the file where the filtered results are stored. (b) Command: cat result-H8 | head -106 | grep sec | awk '{print $3,$5}' > output_H1-H8. Here, "output_H1-H8" is the name of the file where the filtered results are stored.

The filtered results can be checked using the command: more output_H1-H3 or more output_H1-H8.

4 Implementing Load Balancing

As per the steps given in section of implementation of Floodlight Controller, first few steps remains same. After Step 2, load balancing technique is implemented which is demonstrated in the following steps.

Step 1: As discussed earlier in Section, the first step is to run the Floodlight Controller from Floodlight folder. Once the Floodlight Controller is started to avoid any loss of packet, ensure the controller starts sending the LLDP packets from all the enabled ports.

Step 2: In this step of the experiment, Step 2 of section is followed. The command executed is: sudo mn --custom topology.py --topo mytopo --controller=remote, ip=127.0.0.1, port=653. Once this command is executed successfully, check the connectivity between all the hosts using the mininet command: Pingall. Performing pingall as long as there is no packet loss in the network.

Step 3: After successful execution of the above steps, the client and server are defined among the available hosts. Before defining this, check the communication of the hosts over the network with other hosts. The command to perform this is: xterm h1 h1. In the first console of h1, type: ping 10.0.0.3, and in the second console of h1, type: ping 10.0.0.8. When this command is executed successfully, Host-1 is able to communicate with Host-3 and Host-8 over the network.

Step 4: In the next two steps of the experiment, find the best path to Host-3 and Host-8 from Host-1 by observing the traffic on switch-1 ethernet ports s1-eth3 and s1-eth4. To confirm that the packets are being transmitted between the Host-1, Host-3 and Host-8, Wireshark tool is used [19]. (a) Inside Wireshark, click on Capture-> Interfaces and select switch-1 ethernet port-3, i.e., s1-eth3 and start the capture. Now, to filter the packets, in filter section, type: ip.addr==10.0.0.3, where 10.0.0.3 is the ip address of Host-3. It is observed that no packets for Host-1 and Host-3 are sent or received by Host-1. (b) Now, in the filter section, type: ip.addr==10.0.0.8, where 10.0.0.8 is the ip address of Host-8. It is observed that no packets for Host-1 and Host-8 are sent or received by Host-1. Hence, this proves that this is not the current best path.

Step 5: After successful execution of Step 4, check for traffic on switch-1 ethernet port-4. (a) Inside Wireshark, click on Capture->Interfaces and select switch-1 ethernet port-4, i.e., s1-eth4 and start the capture. Now, to filter the packets, in filter section type: ip.addr==10.0.0.3, where 10.0.0.3 is the ip address of Host-3. It is observed that packets for Host-1 and Host-3 are sent or received by Host-1. (b) Now, in the filter section type: ip.addr==10.0.0.8, where 10.0.0.8 is the ip address of Host-8. It is observed that packets for Host-1 and Host-8 are sent or received by Host-1. These steps were followed to find out the current best path for Host-3 and Host-8 communication.

Step 6: The main objective was to create congestion in the network on the best path Host-1->Host-3 and Host-1->Host-8 or vice versa. So, Host-1 pinging Host-3 is enough for the same. In the second console of h1, stop pinging Host-8.

Step 7: In this step of the experiment, we are going to perform load balancing on the network. Load balancing is performed using Dijkstra's algorithm to find multiple paths of same length which will enable us to search for a shortest path in the network. The program written for the same just finds the path with least congestion and forwards traffic on that path. This is done by executing the python script using the command: python loadbalancing.py.

Step 8: Once the load balancing.py script is executed successfully provide input arguments, for this experiment, the testing hosts chosen are: Host-1, Host-3, and Host-8. As discussed earlier, the best among path is to choose for these hosts. The input arguments should be provided as Host-1 as the source and Host-8 as the destination. This can be further interpreted as the first input argument for the script is: 1, i.e., the source host or the client host. The second input argument is: 8, i.e., the destination host or server host Step 9: The load balancer script will update the Floodlight controller's REST API with best path. Once the representational state transfer (REST) application program interface (API) is updated with the best path, the Floodlight Controller is able to communicate the same to the switches in data plane. Thus, application layer plays a vital role in selection of optimum path for controllers in control plane which is carried forwarded by data plane. Once this process is over, the packets passing through the network gets distributed among multiple available paths to perform load balancing. This is how the load balancing technique is implemented.

Step 10: Initially, the optimum path cost will be 0. It is required to run the load balancing script for few times for the statistics to be enabled. Once the statistics are enabled, after some time, the transmission rates will be updated. At this stage, the best path and flows for the best path will be statically pushed to the switches. For checking the flows of the network, perform a REST GET request to: http://127.0.0.1:8080/wm/core/switch/all/flow/json.

Step 11: In Step 6 of the experiment, ping for Host-8 was stopped. Now, since we figured that the best route for Host-1-> Host-8 is found. In second console of h1, type: ping 10.0.0.8.

Step 12: In Wireshark, monitor the interface for switch-1 ethernet port-4, i.e., s1-eth4. (a) In the filter section, type: ip.addr==10.0.0.3, where 10.0.0.3 is the ip address of Host-3. It is observed that packets for Host-1 and Host-3 are sent or received by Host-1. (b) Now, in the filter section type: ip.addr==10.0.0.8, where 10.0.0.8 is the ip address of Host-8. It is observed that no packets for Host-1 and Host-8 are sent or received by Host-1. Thus, after performing the about steps, we can come to the conclusion that this is the best path for Host-1->Host-3 and not Host-1->Host-8.

Step 13: After the successful execution of the above step, now capture the packets on multiple switch ports to check the best route for Host-1->Host-8. Monitor the interface for switch-1 ethernet port-3(s1-eth3), switch-21 ethernet port-1(s21-eth1), switch-21 ethernet port-2(s21-eth2), switch-2 ethernet port-4(s2-eth4), switch-10 ethernet port-3(s10-eth3), switch-18 ethernet port-2(s18-eth2), switch-22 ethernet port-1(s22-eth1), and switch-4 ethernet port-3(s4-eth3). (a) In the filter section type: ip.addr==10.0.0.3, where 10.0.0.3 is the ip address of Host-3. It is observed that no packets for Host-1 and Host-3 are sent or received by Host-1. (b) Now, in the filter section type: ip.addr==10.0.0.8, where 10.0.0.8 is the ip address of host-8. It is observed that packets for Host-1 and Host-8 are sent or received by Host-1. Above-performed steps prove that this is the best path for Host-1->Host-8. And load balancing for network hosts is working.

Step 14: In this step of the experiment, we define one client and two servers from the developed network. The command that is used for this is: xterm h1 h3 h8. This command will open three consoles h1, h3, and h8. To check the configuration details on each of these consoles, type the command: ifconfig.

Step 15: In this step, generate the traffic between client and the servers and record the events using iperf tool. (a) In the h3 console, type the command: iperf -s -p 6653 -i 1 > result2-H3. Here, defining the Host-3 with ip address 10.0.0.3 as the server. Where "result2-H3" is the filename to store the events occurring between Host-1 and Host-3. After this command is executed, the server starts and waits for the client request on the network. Now, in the h1 console, type the command: iperf -c 10.0.0.3 -p 6653 -t 100. Here, for generating the traffic at the client side 10.0.0.1, providing the server ip address 10.0.0.3 and the port number on which the server is listening is 6653. 100 represents time in seconds. (b) Before proceeding with this step, stop the Host-3 server. In the h8 console, type the command: iperf -s -p 6653 -i 1> result2-H8. Now, define the Host-8 with ip address 10.0.0.8 as the server.,where "result2-H8" is the filename to store the events occurring between Host-1 and Host-8. After this command is executed, the server starts and waits for the client request to the network. Now in the h1 console, type the command: iperf -c 10.0.0.8 -p 6653 -t 100. Here, for generating the traffic at the client side 10.0.0.1, provide the server ip address 10.0.0.8 and the port number on which the server is listening is 6653.

Step 16: Next step is to filter specific results obtain from the experiment. From the generated result files in Step 5, further the experiment is continued with these files. For filtering the results grep and awk commands are used for two files as follows: (a) Command: cat result2-H3 | head -106 | grep sec | awk '{print $3,$5}' > output2_H1-H3. Here, "output2_H1-H3" is the name of the file where the filtered results are stored. (b) Command: cat result2-H8 | head -106 | grep sec | awk '{print $3,$5}' > output3_H1-H8. Here, "output2_H1-H8" is the name of the file where the filtered results are stored. The filtered results can be checked using the command: more output2_H1-H3 or more output2_H1-H8.

5 Experimental Issues and Resolution

During the execution of the experiments, researchers came across few issues which were also resolved in due course of time with consistent efforts. The first problem faced was the ….
(a) Floodlight Controller version issue:
Sometimes the Floodlight master version might throw an JSON parsing error, while performing any REST request, so the REST api output might not be getting parsed. To overcome this error, it is suggested to install the Floodlight v1.2 because it supports JSON parsing for performing REST request. This analysis is based on the issues that have been faced while performing REST requests in a network for performing load balancing and has been resolved as provided.
(b) Python package networkx error:
Networkx is a Python package which allows to create, manipulate, and study the network structure. Using this package, the shortest path can be determined for the hosts. To overcome this issue, Python should be installed on the operating system along with that the networkx package is also expected to be installed.

6 Graph Generation

As discussed earlier, final four output files were generated with obtained log before and after load balancing for the given hosts. For the analysis purpose, different network criteria were considered such as packet transfer rate, jitter, and TCP window size. This analysis was performed on the best paths of the network Host-1->Host-3 and Host-1->Host-8. Gnuplot is used for plotting the graphs once the log files are filled with simulation data [20]. The base script for plotting graph is provided in Fig. 4 which is modified as per the requirement in plotting various parameters from log files.

```
set terminal png
set output 'Result1.png'
set xdata time
set timefmt "%s"
set xrange [0.0:200.0]
set xlabel "Interval_Time(In seconds)"
set autoscale
set ylabel "Transfer(In MBytes)"
set format y "%.1f"
set yrange[300:900] writeback
set title " Host-1 to Host-8 Before Load Balancing (BLB) and After
Load Balancing (ALB)"
set grid
set style data linespoints
plot "output_H1-H3" using 1:2 title "Host-1 to Host-8(BLB)",\
"output2_H1-H3" using 1:2  title "Host-1 to Host-8(ALB)" lt rgb
"blue"
```

Fig. 4 Script to generate graphs using Gnuplot

Packet Transfer Size: The first analysis was to measure the packet transfer size for a time interval of 100 s. This is done between Host-1->Host-3 and Host-1->Host-8 for before load balancing and after load balancing, the steps to perform this as follows:

Step 1: The output file for performing this part of the analysis is already generated in the previous sections of this chapter. Our final output files of before load balancing for plotting the graphs are: output_H1-H3 and output_H1-H8. Output files of after load balancing for plotting the graphs are: output2_H1-H3 and output2_H1-H8.

Step 2: Writing two gnuplot files for plotting the graphs for two scenarios Host-1->Host-3 and Host-1->Host->8, where plots.plt and plots1.plt are the file names for the scenarios, whose input files are the one discussed in Step 1. Figure 3 is one of the file used for the same.

Step3: Executing the plots.plt and plots1.plt file using the command: gnuplot plots.plt. At successful execution of the commands, two image files are generated which are the final graph files shown in the next section.

Jitter: The next analysis carried out was jitter rate for a time interval of 100 s. This was again captured between Host-1->Host-3 and Host-1->Host-8 for before load balancing and after load balancing, the steps to perform this as follows:

Step 1: The output files for this part of the analysis were generated separately. We defined one client and two servers from the developed network. The command that is used for this is: xterm h1 h3 h8. This command will open three consoles h1, h3, and h8. This command is executed after the toplology.py script was executed in mininet without any loss of packets in the network. To check the configuration details on each of these consoles, type the command: ifconfig.

Step 2: In this step, generating the traffic between client and the servers before load balancing is performed and recording the events using iperf tool. (a) In the h3 console type the command: iperf -s -p 6653 -u -i 1 > result3-H3. Here, define the

Host-3 with ip address 10.0.0.3 as the server, where "result3-H3" is the filename to store the events occurring between Host-1 and Host-3. "-u" captures only UDP packets instead of TCP. After this command is executed, the server starts and waits for the client request on the network. Now in the h1 console type the command: iperf -c 10.0.0.3 -p 6653 -u -b 10m -t 100. Here, for generating the traffic at the client side 10.0.0.1, providing the server ip address 10.0.0.3 and the port number on which the server is listening is 6653. 100 represents time in seconds. And "-u" allows to capture only UDP packets instead of TCP and "-b" is to set the target bandwidth to 10 m per second. (b) Before proceeding with this step, stop the Host-3 server. In the h8 console, type the command: iperf -s -p 6653 -u -i 1 > result3-H8. Now, define the Host-8 with ip address 10.0.0.8 as the server, where "result3-H8" is the filename to store the events occurring between Host-1 and Host-8. After this command is executed, the server starts and waits for the client. Now in the h1 console, type the command: iperf -c 10.0.0.8 -p 6653 -u -b 10 m -t 100. Here, for generating the traffic at the client side 10.0.0.1, provide the server ip address 10.0.0.8 and the port number on which the server is listening is 6653.

Step 3: In this step, generate the traffic between client and the servers after load balancing is performed and recording the events using iperf tool. (a) In the h3 console, type the command: iperf -s -p 6653 -u -i 1 > result4-H3. Here, define the Host-3 with ip address 10.0.0.3 as the server, where "result4-H3" is the filename to store the events occurring between Host-1 and Host-3. "-u" captures only UDP packets instead of TCP. After this command is executed, the server starts and waits for the client request on the network. Now, in the h1 console, type the command: iperf -c 10.0.0.3 -p 6653 -u -b 10m -t 100. Here, for generating the traffic at the client side 10.0.0.1, provide the server ip address 10.0.0.3 and the port number on which the server is listening is 6653. 100 represents time in seconds. And "-u" allows to capture only UDP packets instead of TCP and "-b" is to set the target bandwidth to 10 m per second. (b) Before proceeding with this step, stop the Host-3 server. In the h8 console, type the command: iperf -s -p 6653 -u -i 1 > result4-H8. Now, define the Host-8 with ip address 10.0.0.8 as the server, where "result5-H8" is the filename to store the events occurring between Host-1 and Host-8. After this command is executed, the server starts and waits for the client. Now, in the h1 console, type the command: iperf -c 10.0.0.8 -p 6653 -u -b 10 m -t 100. Here, for generating the traffic at the client side 10.0.0.1, provide the server ip address 10.0.0.8 and the port number on which the server is listening is 6653.

Step 4: Next step is to filter specific results that are obtained from the experiment. From the generated result files in Step 2 and Step 3, further the experiment is continued with these files. For filtering the results, grep and awk commands are used for two files as follows: (a) Command: cat result3-H3 | head -106 | grep sec | awk '{print $3,$9}' > output3_H1-H3. Here, "output3_H1-H3" is the name of the file where the filtered results are stored. (b) Command: cat result3-H8 | head -106 | grep sec | awk '{print $3,$9}' > output3_H1-H8. Here, "output3_H1-H8" is the name of the file where the filtered results are stored. (c) Command: cat result8-H3 | head -106 | grep sec | awk '{print $3,$9}' > output8_H1-H3. Here, "output4_H1-H3" is the name of the file where the filtered results are stored. (d) Command: cat result4-H8 | head -106

| grep sec | awk '{print $3,$9}' > output4_H1-H8. Here, "output4_H1-H8" is the name of the file where the filtered results are stored.

The filtered results can be checked using the command: more output3_H1-H3 or more output3_H1-H8.

Step5: Now since the files are generated, our final output files of before load balancing for plotting the graphs are: output3_H1-H3 and output3_H1-H8. Output files of after load balancing for plotting the graphs are: output4_H1-H3 and output4_H1-H8.

Step 6: Write two gnuplot files for plotting the graphs for two scenarios Host-1->Host-3 and Host-1->Host->8, where plots.plt and plots1.plt are the file names for the scenarios, whose input files are the one discussed in Step 5. Figure 3 is one of the files used for the same.

Step7: Execute the plots.plt and plots1.plt file using the command: gnuplot plot.plt. At successful execution of the commands, two image files are generated which are the final graph files shown in the next section.

TCP Window Size: In final analysis, TCP window size is measured for a time interval of 100 s. This was again captured between Host-1-> Host-3 and Host-1->Host-8 for before load balancing and after load balancing, the steps to perform this as follows:

Step 1: The output files for this part of the analysis were generated separately. We defined one client and two servers from the developed network. The command that is used for this is: xterm h1 h3 h8. This command will open three consoles h1, h3, and h8. This command is executed after the toplology.py script was executed in mininet without any loss of packets in the network. To check the configuration details on each of these consoles, type the command: ifconfig.

Step 2: In this step, generating the traffic between client and the servers before load balancing is performed and recording the events using iperf tool. (a) In the h3 console, type the command: iperf -s -p 6653 -w 5000 -i 1 > result5-H3. Here, define the Host-3 with ip address 10.0.0.3 as the server, where "result5-H3" is the filename to store the events occurring between Host-1 and Host-3. "-w" is the TCP window size defined for the amount of data that can be buffered without the receiver validating it, in this case the window size is 5000 bytes. After this command is executed, the server starts and waits for the client request on the network. Now, in the h1 console, type the command: iperf -c 10.0.0.3 -p 6653 -w 5000 -t 100. Here, for generating the traffic at the client side 10.0.0.1, providing the server ip address 10.0.0.3 and the port number on which the server is listening is 6653. "-w" is the window size. Here, the TCP window size is considered same for client and server side. 100 represents time in seconds. (b) Before proceeding with this step, stop the Host-3 server. In the h8 console, type the command: iperf -s -p 6653 -w 5000 -i 1 > result5-H8. Now, define the Host-8 with ip address 10.0.0.8 as the server, where "result5-H8" is the filename to store the events occurring between Host-1 and Host-8. "-w" is the TCP window size defined for the amount of data that can be buffered without the receiver validating it, and in this case, the window size is 5000 bytes. After this command is executed, the server starts and waits for the client. Now, in the h1 console, type the command: iperf -c 10.0.0.8 -p 6653 -w 5000 -t 100. Here, for generating the traffic at the client side 10.0.0.1, providing the server ip address 10.0.0.8 and the port number on which the server is listening is 6653. "-w" is the window size. Here the TCP window size is considered same for client and server side. 100 represents time in seconds.

Step 3: In this step, generating the traffic between client and the servers after load balancing is performed and recording the events using iperf tool. (a) In the h3 console type the command: iperf -s -p 6653 -w 5000 -i 1 > result6-H3. Here, defining the host 3 with ip address 10.0.0.3 as the server. Where, "result6-H3" is the filename to store the events occurring between Host-1 and Host-3. "-w" is the TCP window size defined for the amount of data that can be buffered without the receiver validating it, in this case the window size is 5000 bytes. After this command is executed, the server starts and waits for the client. Now in the h1 console type the command: iperf -c 10.0.0.3 -p 6653 -w 5000 -t 100. Here, for generating the traffic at the client side 10.0.0.1, provide the server ip address 10.0.0.3 and the port number on which the server is listening is 6653. "-w" is the window size. Here, the TCP window size is considered same for client and server side. 100 represents time in seconds. (b) Before proceeding with this step, stop the host 3 server. In the h8 console, type the command: iperf -s -p 6653 -w 5000 -i 1 > result6-H8. Now, define the Host-8 with ip address 10.0.0.8 as the server,where, "result6-H8" is the filename to store the events occurring between Host-1 and Host-8. "-w" is the TCP window size defined for the amount of data that can be buffered without the receiver validating it, in this case the window size is 5000 bytes. After this command is executed, the server starts and waits for the client. Now, in the h1 console, type the command: iperf -c 10.0.0.8 -p 6653 -w 5000 -t 100. Here, for generating the traffic at the client side 10.0.0.1, provide the server ip address 10.0.0.8 and the port number on which the server is listening is 6653. "-w" is the window size. Here, the TCP window size is considered same for client and server side. 100 represents time in seconds.

Step 4: Next step is to filter specific results that are obtained from the experiment. From the generated result files in Step 2 and 3, further the experiment is continued with these files. For filtering the results, grep and awk commands are used for two files as follows: (a) Command: cat result5-H3 | head -106 | grep sec | awk '{print $3,$5}' > output5_H1-H3. Here, "output5_H1-H3" is the name of the file where the filtered results are stored. (b) Command: cat result5-H8 | head -106 | grep sec | awk '{print $3,$5}' > output5_H1-H8. Here, "output5_H1-H8" is the name of the file where the filtered results are stored. (c) Command: cat result6-H3 | head -106 | grep sec | awk '{print $3,$5}' > output6_H1-H3. Here, "output6_H1-H3" is the name of the file where the filtered results are stored. (d) Command: cat result6-H8 | head -106 | grep sec | awk '{print $3,$5}' > output6_H1-H8. Here, "output6_H1-H8" is the name of the file where the filtered results are stored.

The filtered results can be checked using the command: more output5_H1-H3 or more output5_H1-H8.

Step5: Now since the files are generated, our final output files of before load balancing for plotting the graphs are: output5_H1-H3 and output5_H1-H8. Output files of after load balancing for plotting the graphs are: output6_H1-H3 and output6_H1-H8.

Step 6: Write two gnuplot files for plotting the graphs for two scenarios Host-1->Host-3 and Host-1->Host->8, where plots.plt and plots1.plt are the file names for the scenarios, whose input files are the one discussed in step 5. Figure 3 is one of the file used for the same.

Step 7: Executing the plots.plt and plots1.plt file using the command: gnuplot plot.plt. At successful execution of the commands, two image files are generated which are the final graph files shown in the next section.

7 Performance Analysis

This section provides the analysis on obtained results obtained from logged records generated at each and every event occurred during every simulation run. The purpose of this experiment was to perform load balancing over software-defined Networks. To test the impact on network after performing load balancing, three most suitable parameters considered are Packet Transfer, Fitter, and Congestion Window.

While referring the scenario provided in Fig. 1, it is clear that the Partially connected topology has been directly been controlled by controller of Floodlight. It can be observed from Fig. 5 which represents Data Transfer between Host-1 and Host-3, that the there is a remarkable increase in data rate of average 85 MB after implementation of load balancing technique for every RTT.

In the similar fashion, Fig. 6 represents data transfer between Host-1 and Host-8 which belong to a far switch where both the main switches are involved for communication. It is observed from the Fig. 6 that in presence of instability, without

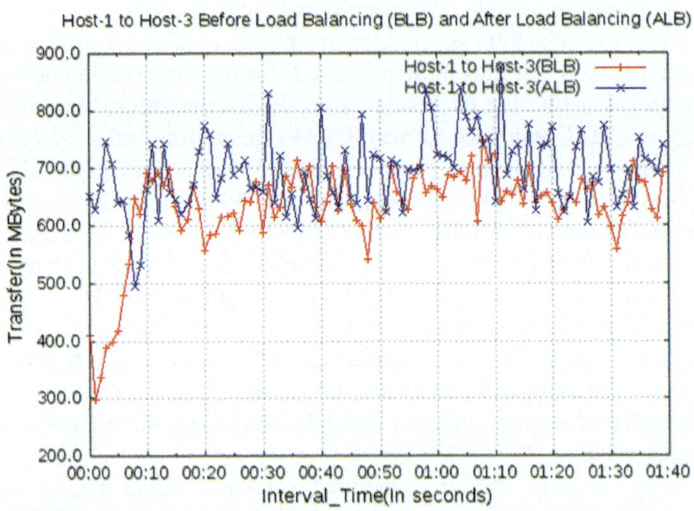

Fig. 5 Host-1 to Host-3 packet transfer before and after load balancing

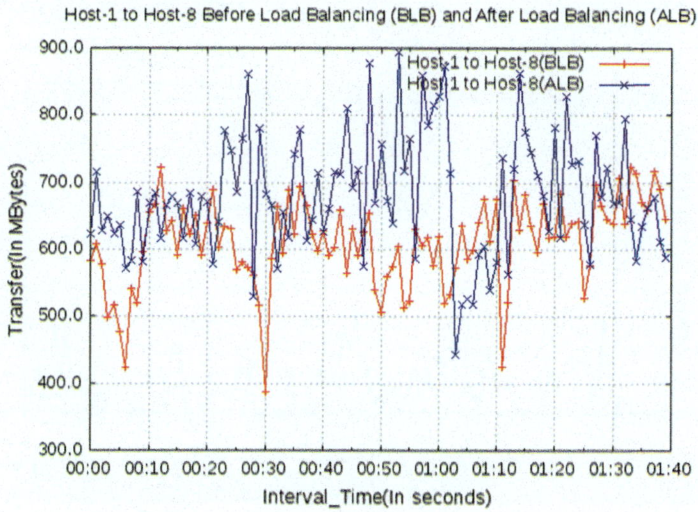

Fig. 6 Host-1 to Host-8 packet transfer before and after load balancing

any event of packet drop remarkable increase of average 123 MB is observed after applying load balancing which is shown in blue line.

To check the load on TCP flows imposed by limitations of network resources, there is a need to observe TCP Window size. More is the window size, less is the congestion existing. Less is the Window Size, more congested is the network. Figure 7 indicates window size of flows observed for 1.5 min in mB. Red line indicates window size

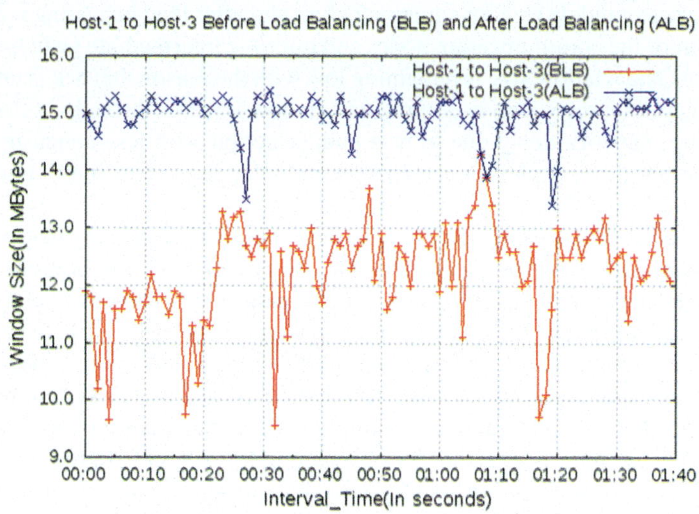

Fig. 7 Host-1 to Host-3 TCP window size before and after load balancing

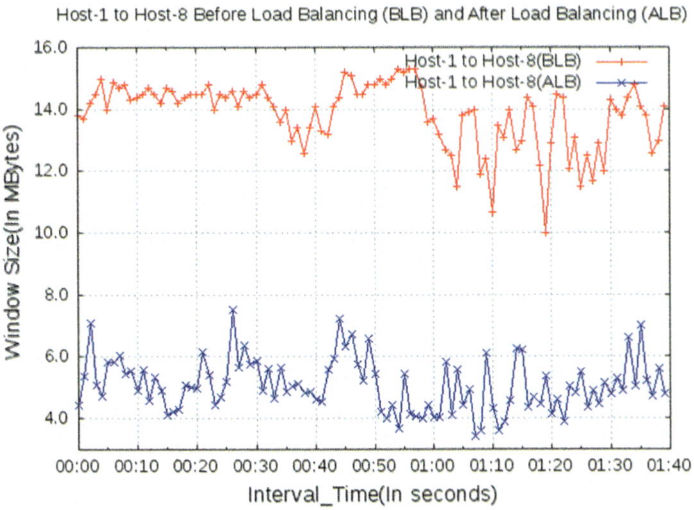

Fig. 8 Host-1 to Host-8 TCP window size before and after load balancing

before applying load balancing while blue line shows after applying load balancing. It is clear that before applying load balancing, the window size is fluctuating from 11 to 13 MB shown with red line points. Whereas, after applying load balancing technique, the window size has become stable throughout the flow life time to 15 MB with only 3 events of decrease in window size.

Considering Fig. 8, it can be observed that the drastic change in the Window Size in comparison of the one observed in Fig. 7. This drastic reduction with a decrease in 7 MB of average window size for situation before and after load balancing is due to the far location of the communicating nodes and multiple intermediate switches coming in between. The window size is remaining low with the reason that ack is taking time in reaching the source which is keeping the window size low throughout the flow life time. Again, network delay due to best path selection with less congestion during load balancing process plays a part in keeping delay high resulting to low window size.

Load balancing techniques may impose delay because complexity lies in selection of alternative path to dilute the clusters that are generated due to bottleneck conditions. To check the load on the system, it is necessary to observe the delay imposed by the Load Balancing Techniques on the system. As a trade off between latency and high data rate, network is performing stable with a rare event of packet drops. Thus, in this experiment we are observing jitter imposed by the system before and after the load balancing techniques are implemented on the network to observe how adversely the network is affected. As observed in Figs. 9 and 10, red line indicates the jitter before load balancing; and blue line indicates jitter after the load balancing in implemented. It can be clearly observed from the graph given in Fig. 9 that only 0.1 ms increase is there in jitter for the flows after load balanced.

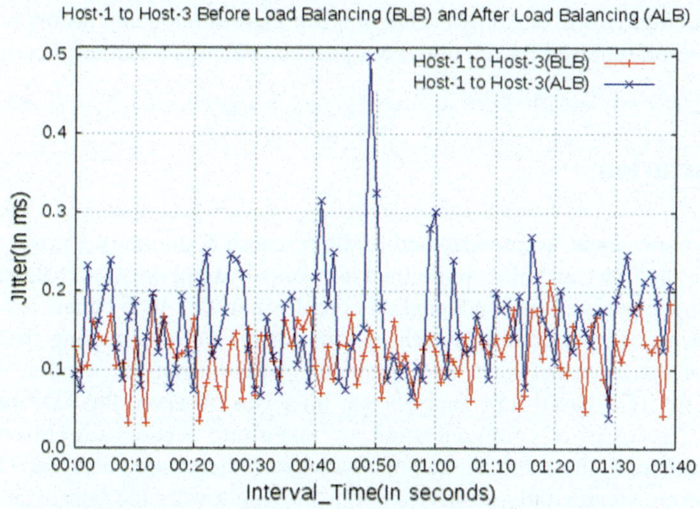

Fig. 9 Host-1 to Host-3 jitter before and after load balancing

Figure 10 indicates the jitter observed for the two communicating host far 3 switches from each other in the situation of before and after load balancing. Observing the graphs given in Figs. 9 and 10 in depth reveals, the situation deteriorates in comparison of that between closely located host given in Fig. 9. But, still the average increase in delay is 0.4 ms which is approximately double of the jitter before applying

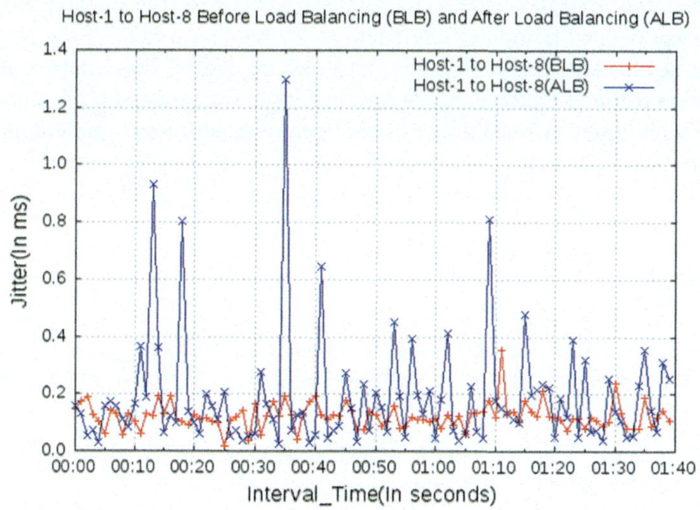

Fig. 10 Host-1 to Host-8 jitter before and after load balancing

load balancing. However, some events of steep increase in delay occurs because of retransmission required for the timed out packets which is a matter of concern.

8 Conclusion

With this experiment, we contributed to the research community a novel approach to manage the flows on software-defined networks by applying load balancing techniques. The experiment is conducted on complex network with connected switched network with complex multiple path options for the packets passing from near to far host of the same network. The parameters of data transfer, window size were observed for TCP-based flow analysis and jitter was observed for UDP-based flow analysis. Two cases of communication are taken into account (a) closely located host (b) far located host and all the three parameters stated are observed. The results obtained were significantly positive and it is moving toward the improvement as far as performance of the software-defined networks is concerned along with reduced load and jitter. The graphs clearly shows green signal to the research community working on SDN load balancing to move ahead considering Floodlight Controller for the base experiments and contribute an improved strategies in the area of load balancing. The experiment also inspires the newbees to explore upon the step-by-step procedure to follow to implement this experiment to set up the base for their research. The chapter will not just set the base of the technological advancement in the area of software-defined networks but will even play a vital role in bringing in a large set of researchers to join the community of SDN research which will result to more contribution and growth of the SDN research. This research chapter will support even not experienced computer engineers to try hands on SDN if the step-by-step procedure is followed with complete focus and no errors. This chapter also opens door for the profound load balancing researchers to explore and try out proved load balancing techniques by extending this research with additional implementations on software-defined networks.

References

1. Senthil Ganesh, N., & Ranjani, S. (2015). Dynamic load balancing using software defined networks. *International Journal of Computer Application, Special Issue*. In *Proceedings of International Conference on Current Trends in Advanced Computing (ICCTAC-2015)* (pp. 11–14).
2. Zhou, Y., Ruan, L., Xiao, L., & Liu, R. (2014). A method for load balancing based on software defined network. *Advanced Science and Technology Letters, 45*, 43–48.
3. Asadollahi, S., Goswami, B., & Sameer, M. (2018). Ryu controller's scalability experiment on software defined networks. In *Proceedings of IEEE International Conference on Current Trends in Advanced Computing (ICCTAC)* (pp. 1–5). IEEE, Bangalore, India.

4. Asadollahi, S., Goswami, B., Raoufy, A. S., & Domingos, H. G. J. (2017). Scalability of software defined network on floodlight controller using OFNet. In *International Conference on Electrical, Electronics, Communication, Computer, and Optimization Techniques (ICEECCOT)* (pp. 1–5). IEEE, Mysore, India.
5. Asadollahi, S., & Goswami, B. (2017). Experimenting with scalability of floodlight controller in software defined networks. In: *International Conference on Electrical, Electronics, Communication, Computer, and Optimization Techniques (ICEECCOT)* (pp. 1–5). IEEE, Mysore, India.
6. Asadollahi, S., & Goswami, B. H. (2017). Revolution in existing network under the influence of software defined network. In *Proceedings of the 11th INDIACom* (pp. 1012–1017). IEEE, New Delhi, India.
7. Goswami, B., & Asadollahi, S. (2017). Implementation of SDN using OpenDayLight controller. *IJIRCCE, 5*(2), 218–227.
8. Manuel, T., & Goswami, B. (2019). Experimenting With scalability beacon controller in software defined network. *International Journal of Recent Technology and Engineering (IJRTE), 7*(6), 1–6.
9. Sameer, M., & Goswami, B. (2018). Experimenting with ONOS scalability on software defined network. *Journal of Advanced Research in Dynamical & Control Systems, 10*(14), 1820–1830.
10. Goswami, B., & Asadollahi, S. S. (2018). Enhancement of LAN infrastructure performance for data center in presence of network security. In Lobiyal, D., Mansotra, V., & Singh, U. (Eds.), *Next-generation networks. Advances in intelligent systems and computing* (vol. 638). Springer, Singapore.
11. Goswami, B., & Asadollahi, S. (2016). Novel approach to improvise congestion control over vehicular ad hoc networks (VANET). In *Proceedings of 3rd International Conference on Computing for Sustainable Global Development (INDIACom)* (pp. 3567–3571). IEEE, New Delhi, India.
12. Goswami, B., & Asadollahi, S. (2016). Performance evaluation of widely implemented congestion control algorithms over diversified networking situations. In *ICCSNIT—2016*, Pattaya, Thailand. Open Access.
13. OpenFlow: FloodLight controller. Available at http://www.projectfloodlight.org/. Last accessed on March 2019.
14. Mininet: Emulator. Available at http://mininet.org/. Last accessed on March 2019.
15. Justin Pettit (2018, August 20). [ovs-announce] Open vSwitch 2.10.0 Available .openvswitch.org. Retrieved March 2019.
16. Python: Scripting network topologies. Available at https://www.python.org/. Last accessed on March 2019.
17. Xterm: Emulator. Available at https://invisible-island.net/xterm/. Last accessed on March 2019.
18. IPERF: Networks tool. Available at https://iperf.fr/. Last accessed on March 2019.
19. Wireshark: Logging and Testing tool. Available at https://www.wireshark.org/. Last accessed on March 2019.
20. Gnuplot: Graph tool. Available at http://www.gnuplot.info/. Last accessed on March 2019.

The manufacturer's authorised representative in the EU is Springer Nature Customer Service Centre GmbH, Europaplatz 3, 69115 Heidelberg, Germany. If you have any concerns regarding our products, please contact ProductSafety@springernature.com

Printed and bound by CPI Group (UK) Ltd, Croydon, CR0 4YY
30/09/2026
02225267-0006